BOGUS
SCIENCE

JOHN GRANT

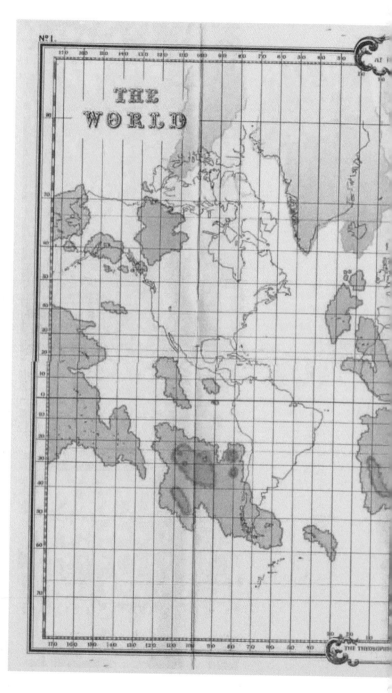

THE
WORLD

BOGUS
SCIENCE

JOHN GRANT

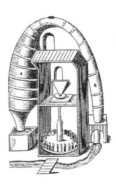

ff&f

BOGUS SCIENCE

Published by Facts, Figures & Fun, an imprint of
AAPPL Artists' and Photographers' Press Ltd.
Church Farm House, Wisley, Surrey GU23 6QL
info@ffnf.co.uk www.ffnf.co.uk info@aappl.com www.aappl.com

Sales and Distribution
UK and export: Turnaround Publisher Services Ltd.
orders@turnaround-uk.com
USA and Canada: Sterling Publishing Inc. sales@sterlingpub.com
Australia & New Zealand: Peribo Pty. michael.coffey@peribo.com.au
South Africa: Trinity Books. trinity@iafrica.com

A catalogue record for this book is available from the British Library.

ISBN 13: 9781904332879

Cover Design: Stefan Nekuda office@nekuda.at
Content Design: Malcolm Couch mal.couch@blueyonder.co.uk

Printed in Malaysia for Imago

Half-title page: See page 256
Title page background: See pages 142–3
Title page: See page 188

CONTENTS

DEDICATION

This book was largely written in the wake
of a rollercoaster ride of surgical operations and
periods of convalescence. The person who bore the
brunt of all this disruption – and who had to
tolerate an often grumpy and frustrated
husband – was my dear wife, Pam.
So, even more than usual,
Thogsbabe, this
one's for
you.

ACKNOWLEDGEMENTS

My profound thanks to

- ❖ Andy Sawyer of the University of Liverpool and the Science Fiction Foundation, who kindly and at short notice provided me with a vital document,
- ❖ Lon Prater, who put me on the track of the Phantom Time Hypothesis,
- ❖ Charles Platt, for giving me permission to quote far more of his article on antigravity research than in the event, for reasons of space, I was able to,
- ❖ The Spammers, who played a major role (as usual) in keeping me sane and who often directed me towards examples of pseudoscientific lunacy,
- ❖ Bill DeSmedt and Fragano Ledgister, likewise,
- ❖ the staff of the West Milford Township Library, who were unfailingly patient when I sought books from afar,
- ❖ Cameron Brown of AAPPL, who gave constant encouragement and who generously permitted me to shift deadlines ever backward,
- ❖ Malcolm Couch, for yet again providing a spiffy design and for being as always a joy to work with,
- ❖ Jane Barnett and Ian Crowther, for being there,
- ❖ Pam, for being *right* there,
- ❖ and all the people I've forgotten I promised to thank.

AUTHOR'S NOTE

———————— ⟨✦⟩ ————————

THIS BOOK FOLLOWS *Discarded Science* (2006), which is primarily concerned with scientific hypotheses – from the woeful to the wonderful – that have fallen by the wayside, and *Corrupted Science* (2007), which examines the ways in which science has been corrupted either by human weakness or more usually by human mendacity, whether grounded in greed, religious belief, bigotry, ideology, politics or any mixture thereof. Both books naturally contain a fair amount about the pseudosciences, especially those related to alien visitors in either the ancient past or, via UFO, the present; but the pseudosciences are not their focus.

In *Bogus Science* the concern is far more with the stuff that walks vaguely like science, quacks vaguely like science, but in fact isn't science at all: it's bogus science, or pseudo-science. This isn't to say that there's not a lot of genuine science within these pages – there is – but it's there in the context of illuminating the bogus.

One thing I realized soon after undertaking *Bogus Science* was that, whereas in the other two books I could have as my aim some approximation, however rough, of compre-hensive coverage of the field, the pseudosciences have today become – in part but only in part because of the internet – so prolific, ubiquitous and many-aspected that I didn't have a hope of succeeding in any kind of quasi-comprehensive approach. Instead I took my inspiration from the title of that 1973 classic *A Random Walk in Science*, compiled by R.L. Weber and edited by R. Mendoza. I decided that for the sake of my own sanity and quite possibly my readers' it was better to concentrate on relatively few areas in some detail than to try to touch every possible base with what would necessarily be infuriating briefness. What you have in your hands, then, is not an

entirely random walk in pseudoscience, but it's quite deliberately a stroll that goes along some lanes and not others.

In particular, I haven't had the space to treat the
psychic/paranormal pseudosciences – from psychometry to
psychokinesis to telepathy to afterlife speculation to astral
travel to reincarnation research to prophecies of the end of
the world*. . . and beyond. It's to be hoped my publisher
will let me make these the subject of a fourth volume –
Spooky Science, perhaps? Likewise, I've largely stayed clear of
bogus medicine and the self-help racket, whether psychically or otherwise based. That, too, is a book in itself.

In the meantime, I hope you enjoy the views from the
lanes down which we *do* have the time and ability to amble.
Any rocks in the road are my fault, and my apologies in
advance for them.

– JG

---

* Or, as my mental editor kept calling these last, "major doomo".

EVERY TIME we let ourselves believe for unworthy reasons, we weaken our powers of self-control, of doubting, of judicially and fairly weighing evidence. We all suffer severely enough from the maintenance and support of false beliefs and the fatally wrong actions which they lead to, and the evil born when one such belief is entertained is great and wide. But a greater and wider evil arises when the credulous character is maintained and supported, when a habit of believing for unworthy reasons is fostered and made permanent. . . . [I]f I let myself believe anything on insufficient evidence, there may be no great harm done by the mere belief; it may be true after all, or I may never have occasion to exhibit it in outward acts. But I cannot help doing this great wrong towards Man, that I make myself credulous. The danger to society is not merely that it should believe wrong things, though that is great enough; but that it should become credulous, and lose the habit of testing things and inquiring into them; for then it must sink back into savagery.

– W.K. Clifford, "The Ethics of Belief" (1877)

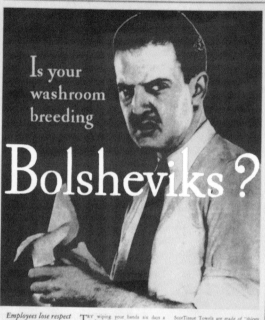

Is your
washroom
breeding

Bolsheviks?

*Employees lose respect
for a company that
fails to provide
decent facilities for
their comfort*

TRY wiping your hands six days a
week on harsh, cheap paper towels
or awkward, unsanitary roller towels—
and maybe you, too, would grumble.

Towel service is just one of those small,
but important courtesies—such as proper
air and lighting—that help build up the
goodwill of your employees.

That's why you'll find clothlike Scot-
Tissue Towels in the washrooms of large
izations such as R.C.A.,
National Lead Co. and
Co.

ScotTissue Towels are made of "thirsty
fibre" . . . an amazing cellulose product
that drinks up moisture 12 times as fast
as ordinary paper towels. They feel soft
and pliant as a linen towel. Yet they're
so strong and tough in texture they won't
crumble or go to pieces . . . even when
they're wet.

And they cost less, too—because one
is enough to dry the hands—instead of
three or four.

Write for free trial carton. Scott Paper
Company, Chester, Pennsylvania.

sue Towels - *really* dry!

Above: Parodying the flood of bogus
information can sell paper towels!

Left: Some people really believe
these things . . .

INTRODUCTION

Of course I know that there will be those skeptics who'll say that this book is all hogwash . . .
    – Sylvia Browne, *Secrets & Mysteries of the World* (2005)

Ignorance is the most delightful science in the world, because it is acquired without pain and keeps the mind from melancholy.
    – Giordano Bruno (1548–1600), *Lo Spaccio de la Bestia Trionfante* (1584; trans as *The Expulsion of the Triumphant Beast*)

There is a difference between having a mind that is open to new ideas and one that is simply vacant.
    – Michael W. Friedlander, *At the Fringes of Science* (1995)

I am often asked why I find it comparatively easy to believe in evaporating black holes and invisible cosmic matter, but not in straightforward things like ghosts and flying saucers that ordinary people apparently see all the time.
    – Paul Davies, "A Window into Science", *Natural History*, July 1993

It's a funny old world out there, isn't it? And a wacky one, too.

In *Mysterious Fires and Lights* (1967) Vincent Gaddis (1913–1997) makes the claim that ball lightning is sentient. Luckily, it seems well disposed towards human beings – in fact, according to Gaddis, it can on occasion be "socially minded". As evidence, he cites some of the examples described by the French astronomer Camille Flammarion (1842–1925) in which domesticated animals were killed by exploding ball lightning while humans in closer proximity to the explosion were left unharmed.

In *Sacred Science: The King of Pharaonic Theocracy* (1961) René Schwaller de Lubicz (1887–1961) spells out some important horticultural mysteries:

If a good gardener plants his cauliflower on the day of the full moon, and a bad gardener plants them at new moon, the former will have rich, white cauliflower and the latter will harvest nothing but stunted plants. It is sufficient to try this in order to prove it. *So it is for everything that grows and lives.* Why these effects? Direct rays of sunlight or indirect rays reflected from the moon? Certainly, but for quite another, less material reason: *cosmic harmony.* Purely material reasons no longer explain why the season, even the month and the precise date, must be taken into account for the best results. Invisible cosmic influences come into play . . .

. . . while alternatively, if The Cauliflower Code is insufficiently fringe, the shelves of your local bookshop are likely to be full of titles that promise unusual extensions to the knowledge you might recognize from *New Scientist* or *Scientific American*: books like Donald L. Wilson's pioneering *Natural Bust Enlargement with Total Mind Power: How to Use the Other 90% of Your Mind to Increase the Size of Your Breasts* (1979) and Gary Leon Hill's useful instructional text *People Who Don't Know They're Dead: How They Attach Themselves to Unsuspecting Bystanders and What to Do About It* (2005).*

So far as many people are concerned, the spread of bogus science throughout our society – from Creationism to belief in UFO abductions – isn't anything to be troubled by: it's merely a matter of amusement. And, of course, no one wants to be seen to be questioning the principle of free speech.

The US, the Western nation most seriously disabled by the widespread promulgation of nonsense, has recently received some hard blows to its self-image as the best-educated nation in the world. The 3rd International Mathematics and Science Study, released in 1998 by the International Association for the Evaluation of Educational achievement, examined kids aged 17–18 in 23 countries. In physics the US youngsters came last. In advanced math, they came second-last. In maths/science they came fourth last. In other words, US kids seem to have the levels of knowledge and understanding that might shame many an

---

* What seems symptomatic of such books is the length of their subtitles!

impoverished Third World country – Cuba, for example. Various explanations have been offered for this, but the most reasonable is the way that, in a perversion of the original notion of free speech, far too many of us now regard *everything* as being open to debate and rival interpretation, with the very nuttiest and least reality-based of those alternative interpretations being given the same weight as others of genuine worth. Thus the doctrine of Intelligent Design, which is really just Creationism slathered in a pompous mud, is treated as if it were a serious contender to the well established theory of evolution by natural selection.*

We like to kid ourselves that our society is becoming progressively better educated. In his book *Behind the Crystal Ball* (1996) Anthony Aveni discusses some historical polls about science, magic and bogus science.

When Columbia University students were polled in 1920, just 2% believed there was anything in palmistry and just 4% believed in astrology, but about 10% credited phrenology, avoided the number 13, and thought you could make someone turn to look at you if you stared hard enough at their back (not necessarily the same 10% in all three cases).

By the 1950s in the UK, 20% credited astrology, 17% believed in ghosts (and about 7% believed they'd seen one), and over 15% swore by such superstitions as lucky numbers and mascots. In Germany at roughly the same time, about one-third of the people believed in astrology while the levels of superstition were substantially higher than the UK equivalents.

In a 1977 Roper survey in the US, belief in astrology ran at 25%, ESP 53%, Heaven and Hell 74%, UFOs as alien artefacts 29%, and reincarnation 14% – this last figure was down from a 1969 survey which had shown belief in reincarnation running at 20%. Another US survey the following year put belief in ghosts and witches at about 10%, angels at 54%, flying saucers at 57% and the Devil at 37%, with credence given to astrology by nearly 50% and to precogni-

---

* The fact that a lot of people don't know the meaning of the word "theory", a failing deliberately encouraged by many in the Creationist camp, obviously doesn't help.

BIKAMASUTRAL MIND?

In *Right Brain Sex: Using Creative Visualization to Enhance Sexual Pleasure* (1989) Carol G. Wells went places even Julian Jaynes, author of the ground-breaking *The Origins of Consciousness in the Breakdown of the Bicameral Mind* (1976), never imagined. Here's a selection of her headings and subheadings. Note how this differs from a paper in *Nature*.

- ❖ How Visualization Bypasses Your Sexual Roadblocks
- ❖ How Using Your Right Brain Makes You Passionate and Overcomes Boredom
- ❖ How Bored Are You with Your Sex Life?
- ❖ Are You Having Trouble Concentrating during Sex? Techniques That Help
- ❖ Are Your Notions about Masturbation Part of the Problem or Part of the Solution?
- ❖ A Historical Look at Masturbation
- ❖ What Lust Does for Your Sex Life
- ❖ The Mutual Exclusiveness of Guilt and Lust
- ❖ Pleasure: The Final Destination
- ❖ Are You in Tune with Your Body?
- ❖ Can You Surrender to Pleasure?
- ❖ How Vulnerable Are You? A Personal Test
- ❖ Power and Pleasure – The Oil and Vinegar of Great Sex
- ❖ Orgasms Too Soon, Orgasms Too Late, or Possibly Erections That Won't Cooperate – Are They Possible to Eliminate?

Someone, somewhere, should take the time to set that last one to music.

tion by over one-third. By 1981 the belief in reincarnation had risen to 23%.

According to a 1987 report in *Time*, whereas only one-half the US population had believed in psychic phenomena in 1974, now (i.e., in 1987) the figure was more like two-thirds. Of those surveyed, 15% claimed to have seen a flying saucer, 62% believed in the Devil (although "only" 54% in demonic possession) and 25% believed in astrology. In 1993, 69% believed in angels and 49% in devils. Belief in ESP ran at 46%, in clairvoyance at 22%, and in communication with the spirits of the dead at 14%.

Even in the 21st century, only about 10% of US adults know what radiation is, about 30% know that DNA is the key to heredity, almost nobody outside scientists knows what a molecule is, and an astonishing 20% think the sun goes round the earth. And 40% of US *scientists* think they can communicate directly with God. A US survey done by Gallup in 2009 to mark the 200th anniversary of Charles Darwin's birth revealed that a full 25% of Americans rejected evolution outright, an extraordinary 36% thought the matter was still open to debate and "don't have an opinion either way", while only 39% accept the reality. Meanwhile, a UK survey published in 2009 showed how much more credulous the Brits have grown since the 1950s: 39% believe in ghosts, 22% in astrology, 27% in reincarnation, 53% in life after death and 55% in Heaven. (The 2% difference in the latter is puzzling: presumably a few people think we don't survive death but might go to Heaven anyway.)

These figures do not paint a progressively more educated society. It's difficult to regard them as anything other than a massive failure in our public education, which in itself reflects a massive failure of responsibility by our politicians and media. But they also represent a failure by *us*, because each of us individually is almost certainly not doing enough to beat back the flood of bogus science. And there's probably not one of us who hasn't on occasion been fooled by it.

According to the paper "Learning of Content Knowledge and Scientific Reasoning Ability: A Cross Cultural Comparison" by Lei Bao *et al.*, published in *Science* in January 2009, which compared US college scientific and engineering freshmen with their Chinese counterparts, the ability to reason scientifically is at a low ebb in both cultures, with the US students being significantly more ignorant than the Chinese ones about matters scientific.

Both groups did poorly in the test of basic knowledge about electricity and magnetism, but the difference in quality of school science education between the two countries is dramatically reflected in the average score of the two groups for this test: the Chinese averaged just under 66%, the US students just under 27%. (As the paper's lead author, Lei Bao of the Physics Education Research Group at Ohio State University, pointed out, the US students' average of 27% is not too significantly above the score that could have been achieved by chance in this test, 20%.) In the test of knowledge of mechanics, both groups did better, but the contrast between average scores was still major: 86% for the Chinese and 49% for the Americans.

Yet in their ability to reason scientifically, as indicated by the Lawson Classroom Test,* the average scores for the two groups represented a statistical dead heat: just under 75% for the Chinese and about 74% for the Americans. Before the latter think of getting cocky, though, it should be noted that 75% is regarded as a fairly poor score in this particular test.

The counterintuitive subtext here is that mere knowledge of scientific facts does not much affect the ability to reason logically. Or maybe it's not so counterintuitive after all: as a few voices in the wilderness have been saying for decades now, a major flaw in almost all modern education systems is that they assume the *ability to think* is somehow inherited or is a natural property of the human brain, rather than something that needs to be learned – and taught. In other words, we're not really equipping our young with the ability to recognize *why* the bogus scientific claims presented to them from all directions in our society are indeed bogus; we're just *telling* them those claims are bogus because they "disobey the rules" – never the best way to encourage people, especially young people, to avoid something!

* A patented scheme whereby students are presented with scientific hypotheses and asked to evaluate them using deductive reasoning.

> The sum total of food converted into thought by women can never equal the sum total of food turned into thought by men. It follows, therefore, that *men will always think more than women.*
>
> – Miss M.A. Hardaker,
> "Science & the Woman Question",
> *Popular Science Monthly*, 1882

If at the moment the ones who're suffering most from the failure to educate scientifically are the young, who're expected to compete in a world for which their education has woefully ill equipped them, in the near future it'll be their elders – the people who enabled this abysmal situation to arise – who also suffer, as the US economy reels under the consequences of years of tolerating the inability to differentiate between nonsense and reality.

That's even before we start to consider the potentially genocidal effect of the bogus science deployed by very powerful figures to promote opposition to warnings of the lethal and imminent dangers of climate change. In this context, with perhaps billions of lives at stake, how sensible is it for a culture to regard the promotion of bogus science as a freedom of speech issue?

These are very difficult problems to address. One possible partial answer might be the encouragement of a more responsible attitude among the media – from TV programmes to websites to books and newspapers – such that the current false notion of journalistic balance (which thinks impartiality is giving equal credit to an expert and a fruit-bat) be replaced by a more genuine balance of treatment in which rational arguments are portrayed as such and the lunatic fringe likewise. If something like this – some way of demarcating bogus science from the rational – doesn't happen soon, the consequences are likely to be irremediable.

It's ironic, in light of the level of technology surrounding us in our daily lives, that we live in such an unscientific or even antiscientific age. To say there are Luddites loose would be to misrepresent matters: what motivated the Luddites was that they *did* understand the technology they hated – all too well. The same excuse cannot be made for the Creationists, anti-evolutionists and others who both loathe science and are – often wilfully – uninformed of its basics. Not content with wallowing in their own ignorance, many actively crusade to persuade others likewise to turn their backs on science, using not just blatant proselytizing but also various quite consciously deceptive means.

A case in point is the novel *The Darwin Conspiracy* (1995) by James Scott Bell,* whose conceit is that a multiple murderer and (even worse) atheist called Sir Max Busby, inspired by a personal detestation of God,† sets out to defile and degrade society by leading people away from a Christian fundamentalist reading of the Bible. In order to do so, Busby plants the theory of evolution by natural selection in Darwin's mind, promotes it with the assistance of Sir Charles Lyell and T.H. Huxley, murders Darwin, and during the rest of his inordinately long (but astonishingly celibate) life spreads his creed far and wide, convincing movers and shakers like Bertrand Russell, Adolf Hitler, Margaret Sanger, Karl Marx and any other contemporaneous figure Bell doesn't like. The narrative is amplified by supposedly nonfictional endnotes in which Bell believes he's making killing jabs at science; in fact, they come across like the uninformed barbs of a bratty adolescent who isn't half as clever as he thinks he is.

The narrative itself is full of the type of antiscientific illogic that wouldn't fool a bright five-year-old but seems intended to fool Bell's audience, as in this conversation between narrator Sir Max and Clarence Darrow:

* Not to be confused with the novel *The Darwin Conspiracy* (2005) by John Darnton, which is a wonderfully enjoyably romp, or *The Darwin Conspiracy: Origins of a Scientific Crime* (2008) by Ray Davies, nonfiction claiming Darwin plagiarized from various scientific contemporaries. Both books are far crueller to Darwin's reputation than is Bell, yet neither draws the kind of scientific opprobrium his does. Advocates of ID might ask themselves why this is so.

† In which case he couldn't be an atheist, surely?

> I perked up. "Are you saying that man has no free will?"
>   "And why should that be so surprising? Hasn't Darwin taught us that we are basically machines?"
>   "Why yes, Darwin has taught us something like that."
>   "And what will does a machine have?" . . .

Of course, Darwin didn't teach us that we're "basically machines", or anything of the sort. Even had he done so, the leap from there to a claim that this obviates human free will is not logical. After a while one can't help sympathizing with supposed villain Joel Nairobi, who finds himself in a situation all of us recognize if we've ever debated a Creationist:

> "Do you know what the odds are against the basic enzymes of life arising from chance? About one in ten to the forty thousandth power. A mathematical impossibility. With these odds, I put my money on an intelligent designer. What do you think?"
>   "I think," said Nairobi, "that I am getting a headache."

An early study of bogus science was *Wish and Wisdom: Episodes in the Vagaries of Belief* (1935) by the psychologist Joseph Jastrow (1863–1944); the same author's *Fact and Fable in Psychology* (1900) had some overlapping material. At the beginning of the 1935 book Jastrow spells out what he describes as the Seven Inclinations, whereby otherwise perfectly intelligent, rational people come to believe purest hogwash. His "inclinations" are:

- ❖ **Credulity** – or gullibility
- ❖ **Marvel** – the urge to accept magic, which overwhelms us when we're in infancy and against which, in later life, rationalism can sometimes wage a losing struggle
- ❖ **Transcendence** – the belief in powers that transcend the natural
- ❖ **Prepossession** – the mental phenomenon whereby, when we seek evidence of our preconceptions, we find it
- ❖ **Congeniality of Conclusion** – whereby we reach the conclusion we *like* rather than the one dictated by evidence and logic

❖ **Vagary** – the obsessive pursuit of a particular conclusion, decided upon early, whatever the contrary evidence

❖ **Rationalization** – the intellectual art of piecing together valid evidence in such a way as to produce an invalid conclusion

One further piece of bogus thinking that turns up again and again among the pseudoscientists and their cohorts is the so-called god-of-the-gaps fallacy. This is the line of reasoning whereby, if science has yet to come up with an accepted rational explanation for a phenomenon, then the only possible explanation must be the one which the bogus thinker has advanced.

The fallacy got its name because the argument used to be used by the Church to justify its "explanation" of such phenomena as the origin of life: if science didn't know the details then God must have been responsible for that first vital spark. Likewise, if science didn't know why fossils of sea creatures were being found on mountaintops, the only possible conclusion must be that the Bible's story of the Flood is literally true.

Of course, the fallacy relies upon false either/or juxtapositions, and most theologians (and believers in general) have moved on to more sophisticated reasoning (alas, not all). But it's alive and well in the works of the pseudoscientists. Read any book by Erich von Däniken, to choose one example among countless, and you'll come across numerous statements of the type "There can be no other explanation for . . ." Of course, in each instance there are myriad alternative explanations to the one von Däniken is proposing for some supposed archaeological mystery, the most likely usually being that he's got his facts wrong, but the attempted logical legerdemain is the same as the old "if you cannot account for this otherwise, God must take the credit".

Again, we're back to the matter of the importance of learning how to think rationally.

Modern cranks are fond of chorusing to the effect that "They laughed at Galileo, they laughed at Pasteur, they laughed at . . ." – the implication being, of course, that in due course science is likely to have to come to accept Bloggs's theory that the universe is a giant newt, just as science came to accept Galileo's support for the Copernican theory and his observations of the moon, Venus and the Jovian moons, and Pasteur's germ theory of disease and demolition of the theory of spontaneous generation (the notion that the smallest living creatures – microbes – were generated from inorganic material).

The problem for Bloggs's supporters is that in fact they – meaning the scientific establishment – *didn't* laugh at Galileo, Pasteur and other, similar paradigm-shifters. Galileo was persecuted by the religious authorities, not the scientific ones; he had the respect, as a scientist, of his peers.* Those scientists who regarded the astronomical phenomena he observed as likely the product of flaws in his telescope were not being unreasonable: his telescope was an extremely primitive affair and its lenses lousy, and many of the things seen through it – such as colour fringes – were indeed the product of optical flaws. In Galileo's day not enough was known about optics for him to be able to explain why observers should take some of the things they saw through his telescope as genuine and discard others as instrumental artefacts. It was perfectly rational for his contemporaries to reserve judgement; and even those who disagreed with him continued to regard him as a major scientific figure.

Similarly for Pasteur. The French Academy of Sciences very promptly verified his work disproving spontaneous generation; those who mocked him were elements of the popular press. When he produced his germ theory of disease, the medical establishment rightly declined to accept it until he produced some pretty strong experimental proof. Once he'd done that, the theory was soon embraced. And again he was never regarded by the scien-

---

* Although often the respect was grudging. In his social interactions Galileo was a pain.

tific world as anything other than an important researcher and theorist.

This is not the case with Bloggs and his newt theory.

The best example any Bloggsian might hope to produce is that of Alfred Wegener and his long-rejected hypothesis of continental drift; this finally came to be accepted by the majority of earth scientists long after Wegener's death, when the phenomenon of seafloor spreading was discovered and thereby a mechanism revealed for the drift. Wegener's problem had been that he couldn't exhibit any such mechanism beyond a sort of woffly appeal to a hitherto unknown force called *Pohlflucht*. Further, he was a meteorologist, not an earth scientist; in terms of the earth sciences he was as much of an amateur as you or I. Even so, during the decades between Wegener's enunciation of the hypothesis and its eventual confirmation, a respectable minority of earth scientists *did* accept his ideas; the rejection was by no means universal, and Wegener was not regarded as a crackpot.

The point is that, *even at the time*, it's generally fairly easy to tell the difference between a potential paradigm-shifter and a crank. Potential paradigm-shifters work within the field in which they're producing their speculations, or at least within a closely related one; they produce research work in order to support their hypotheses; they publish their results and their hypotheses in such a way that these may be reviewed and criticized by their scientific peers – rather than, like Bloggs, getting a fat advance out of a publisher. Failing that, for Bloggs, there's always the internet . . .

The scientific establishment is not stupid to resist radically new hypotheses; if science happily accepted every new notion that came by, human knowledge would rapidly become a complete shambles. Superficially appealing hypotheses have to be adequately tested and debated to make sure they're not subtly nonsensical before they can be added to the corpus of scientific understanding. Once Hypothesis A has achieved this, then science is similarly wise not to discard it again in favour of Hypothesis B without first making sure that A is indeed flawed or incomplete. The sluggishness of science to alter its stance may seem

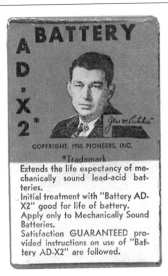

Jess M. Ritchie
personally promoted
his own battery
additive, AD–X2

exasperating from the outside, or to those with radical ideas
that are worthwhile; but in fact it's a remarkably useful characteristic.

Here's an object lesson for us all concerning the day-to-day
practical costs of laziness in resisting bogus science and of
scientific illiteracy.

In the aftermath of WWII the US suffered a shortage of
lead, and this affected (among much else) the manufacture
of car batteries. The search was on for additives that might
extend battery life, and sure enough a number of these
appeared on the market: Bat-Re-Nu and Duble Power were
two. The National Bureau of Standards (NBS) tested all of
these and declared them at best worthless; the test results
were circulated by the National Better Business Bureau and
others, and the products fell into disuse.

All but one, AD–X2. The proprietor of AD–X2, one Jess
M. Ritchie, in 1947 declined to patent the substance, so had
no requirement to state what was actually in it. He did,
however, just a whisker ahead of the Better Business Bureau,

request that the NBS test it. The NBS responded that it did not carry out its testing at the behest of commercial enterprises. Ritchie gathered support from California Senator William Knowland and from the Oakland Chamber of Commerce. Finally, though, it was the Federal Trade Commission that persuaded the NBS to investigate AD–X2 – which the NBS did, finding it valueless. The Federal Trade Commission asked the NBS to test the product again; again the NBS's scientists found it useless.

In 1952 the US Post Office began an on-again-off-again policy whereby sometimes Ritchie was told he could not use the mails to further what was evidently a fraudulent business. Ritchie appealed to the Senate and House Committee on Small Business, who required the NBS to test AD–X2 *yet again*! The results of the NBS's testing showed . . . but you're way ahead of me, aren't you?

So one of the Select Committee's staffers asked MIT to run its own test on AD–X2. The MIT report was a mite less condemnatory than the NBS reports had been, but still found no reason to suspect AD–X2 might actually be useful. This was described by some unknown on the Select Committee or its staff as somewhat favourable . . . and so the saga continued.

In 1953 the Eisenhower Administration arrived in the White House, and with it Sinclair Weeks (1893–1972) as Secretary of Commerce; two of his first actions were to fire the head of the NBS and to suspend the publication of the NBS's bulletins concerning the efficacy or otherwise of battery additives! The National Academy of Sciences protested, and so – in what reads at this distance in time as an instance of mere gesture politics – Weeks asked the Academy to perform its own tests of AD–X2. To no one's surprise, the NAS found that AD–X2 was indeed worthless, just as the NBS had by now several times reported. A few months later the fired NBS head was reinstated.

That should have been the end of it, but of course it wasn't. The Select Committee continued to debate the issue on the grounds that

> Political decisions and policies are sometimes found necessary to mediate, postpone or circumvent the effect of harsh and

arbitrary findings of science that impose unacceptable obliga-
tions or conditions on the electorate or the individual. . . .
[T]he primary role of science in commerce should not be to
regulate the quality of products to protect the consumer but to
discover the truths of nature, and use them more particularly
to create additional products for human satisfaction and entrepreneurial exploitation.*

At one point matters got so out of hand that the Justice
Department was instructed to prepare an anti-trust suit
against the Association of American Battery Manufacturers,
who'd been involved in the case from an early stage on the
perfectly reasonable basis that problems associated with
useless additives could well redound on its members.

Over a period of many years, then, millions of taxpayer
dollars were wasted on a substance that every scientific test
said was useless for its purported function, as if further
debate might somehow alter the results of the scientific
tests. This is not rational thinking. Reality does not bend to
our whims.

In succeeding decades the Pentagon was going to
demonstrate with a vengeance, over and over again, the
inordinate waste of resources, taxpayer monies and human
energies that invariably follow when lay people with axes to
grind permit their preconceptions to overrule the findings
of science and support the claims of bogus science. Just
think of the hafnium bomb. Or Star Wars.†

It's at about this moment that you might want to grab your
comfort-blanket copy of *How to Have an Out-of-Body
Experience in Thirty Days* (1989) by Keith Harary and Pamela
Weintraub – published not by some obscure New Age outfit
in a California burg you've never heard of but by St Martin's
Press in New York. Alternatively, read on . . .

---

* Staff report in *Technical Information for Congress: Report on the
Subcommittee on Science, Research and Development of the Committee on
Science and Astronautics*, 1969, 4.

† For more on both, see my book *Corrupted Science* (2007).

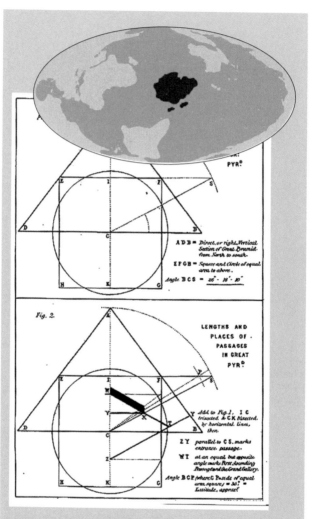

Top: See page 136
Above: See page 151

BOGUS SCIENCE

or

Some People Really Believe These Things

> "It is easy to see how such a picture could fool the untrained eye."
> – Spokesman for the International Flat Earth Research Society, reacting to the first satellite photos of the earth.

Some people really believe the earth is flat.

Almost equally untenable is the claim that belief in the flat earth has been only an occasional, generally short-term player in human history – more specifically, that the Christian Church has never maintained that we live on a two-dimensional world. The truth is that the cosmology of the Bible is both geocentric (earth is at the centre of the universe) and planist (earth is flat); this is hardly surprising, given that all the Middle East cosmologies at the time the Bible was being written were likewise planist. The Hebrew cosmology of the Bible was derived from that of the Babylonians, in which the flat earth has a physical dome overarching it, with waters all around – both beyond the dome and beneath the earth.* Also below the earth was Sheol, the Hebrew equivalent of Hades. Egyptian, Sumerian and Hebrew cosmologies differed in their details – sometimes quite major details – from the Babylonian model, but their planism was emphatically shared. Thus in *Genesis* we find:

> 1:6 And God said, Let there be a firmament in the midst of the waters, and let it divide the waters from the waters.
> 1:7 And God made the firmament, and divided the waters which were under the firmament from the waters which were above the firmament: and it was so.

---

* What we call stars and planets were, to the Babylonians, gods and goddesses; no wonder astrology came into being, because clearly the movements and dispositions of these celestial beings must be of paramount importance to events on earth.

Views hadn't changed much by the time of the New Testament, as we discover in *Matthew*:

> 4:8 Again, the devil taketh him up into an exceeding high mountain, and sheweth him all the kingdoms of the world, and the glory of them;
> 4:9 And saith unto him, All these things will I give thee, if thou wilt fall down and worship me.

Leaving aside such concerns as atmospheric refraction, "all the kingdoms of the world" might indeed be visible from a high enough mountain if the world were flat; clearly on a spherical earth the trick is impossible.*

Countless further examples could be offered of planist statements in the Bible, and indeed those statements *were* liberally offered to the faithful by the Christian Church until about the 6th century. Even in the time of Christ, spherical (i.e., earth-is-spherical) cosmologies had been becoming generally more accepted, whatever the Bible might say; there was argument among the early Christians about whether they should believe the Bible or the cutting-edge science of their day, and naturally many churchmen (originally the overwhelming majority of them) chose the former and attempted to impose a similar choice upon their flock. Tertullian (*c*160–*c*220), Lactantius (*fl*313) and Theodore of Mopsuestia (*c*350–428) were major Christian figures who insisted upon the planist cosmology; a later Christian champion of planism was Cosmas Indicopleustes (*fl*550).

According to Cosmas's *Christian Topography* (*c*550), the earth was a rectangle of extent 400 days' journeying east–west and 200 days' journeying north–south. Angels were responsible for the motions of the celestial bodies. Rain was the product of waters falling in through the heavenly vault. The sun must be near and small because, for example, Isaiah was able to persuade the Lord to move the sun backwards temporarily in its course as a sign to the ailing Hesekiah that he would indeed recover (*4 Kings* 20:8–20:11).

---

* The earth is, strictly speaking, an oblate spheroid. Here the terms "spherical" and "sphere" are used loosely for the sake of ease.

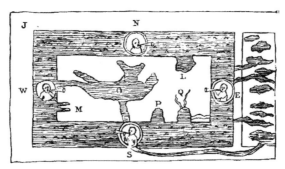

Cosmas's map of the flat, rectangular world described in the Old Testament

By the time Cosmas was writing, however, the planist argument had essentially been lost, within the Church as much as outside it.* In many ways it's surprising the argument *was* lost, because belief in a globular earth required the Church to at the very least apply some creative "interpretation" to the many passages of the Bible that depict terrestrial flatness. (It is equally surprising that relatively few modern self-proclaimed Christian Fundamentalists, with their stated belief in the literal truth of every word in the Bible, are planists.) The process of shifting the clerics to an essentially unanimous acceptance of the spherical cosmology took until about the 8th century, and by the later stages of the Middle Ages this acceptance was universal among the European educated classes (and, very likely, widespread among the rest of society). Thus the classic tales of Columbus having to persuade the Spanish court the earth was round and later of his facing mutiny by crewmen fearing they'd sail off the edge of the world are mere legend, invented by Washington Irving (1783–1859) for his *History of the Life and Voyages of Christopher Columbus* (1828).†

* Not so the geocentric argument, of course: that particular fallacy hung around in the Church for several centuries more, culminating most notoriously in *l'affaire* Galileo.

† The modern flat-earther Charles Johnson (see page 54) claimed that the midocean fracas between Columbus and his men arose because

Whatever the shape of the earth, religious dogma continued for centuries to dominate speculation as to the earth's nature. For example, in his *Telluris Theoria Sacra* (1681; vt *The Sacred Theory of the Earth*) the English theologian Thomas Burnet (c1635–1715) put forward a theory of geomorphology whereby the nascent earth, home to Adam and Eve, was a perfectly smooth hollow sphere, unaffected by seasons, its surface unmarked by mountains or seas, all of the waters being inside the shell. Then:

> . . . as the Heat of the Sun gave force to these Vapours more and more, and made them more strong and violent; so the other Hand, it also weakened more and more the Arch of the Earth that was to resist them, sucking out the Moisture that was the Cement of its parts, drying it immoderately and chapping it in sundry Places. And there being no Winter then to close up and unite its Parts, and restore the Earth to its former Strength and Compactness, yet grew more and more disposed to a Dissolution. And at length, these Preparations in Nature being made on either side, the Force of the Vapours increased, and the walls weakened which should have kept them in, when the appointed time was come, that All-Wise Providence had design'd for the Punishment of a sinful World, the whole Fabrick brake, and the Frame of the Earth was torn in Pieces, as if by an Earthquake; and those great Portions or Fragments, in which it was divided, fell down into the Abyss, some in one Posture, some in another . . . and the Parts that stood above the Waters are the Mountains and Precipices that we admire today.

To us this might seem like mythology, but to Burnet and his contemporaries it represented a welcome scientific approach to geology. Among Burnet's supporters was Isaac Newton (1642–1727), who sent him the suggestion that days might have been longer when the earth was young; this notion Burnet rejected as unscientific, on the grounds that it'd have required a special intervention by God to shorten the days to the length they are now.

In his later book *Archaeologiae Philosophicae: Sive Doctrina Antiqua de Reum Originibus* (1692) Burnet made a

Columbus knew perfectly well that the world was flat whereas his sailors, believing in a globular earth, were terrified of going far enough around the globe's curve that they'd not be able to get back.

further proposal that was simply too scientific for the theologians of his time, who succeeded in having him dismissed from the court of William III. This proposal was that the story of the Fall of Man might be a symbolic rather than a literal one. Three centuries later, the same sort of thing can happen to schoolteachers for the same sort of reasons.

Most of the flat-earth crusaders over the succeeding centuries have been religious in their inspiration. One exception was Sir Richard Phillips (1767–1840), whose motives in promoting planism were so far divorced from religious in nature that he served time in prison for disseminating atheist literature!

It wasn't until the second half of the 19th century that the next major planist revival came. This was largely thanks to the indefatigable work of quack and Biblical literalist Samuel Birley Rowbotham (1816–1884), who for his planist lectures and writings went under the pseudonym Parallax. His *magnum opus* was *Zetetic Astronomy* (1881), over 400 pages of tightly argued reasoning including not just arguments from theological bases but appeals to science; Parallax used science much as do modern adherents of Creationism/ID, picking and choosing his evidences and trying to persuade his audience according to the god-of-the-gaps fallacy (see page 20): that a problematic result in orthodox science, however trivial, is necessarily evidence in favour of whichever baloney the arguer believes in. A further element of *Zetetic Astronomy* comprised experiments that Parallax claimed himself to have carried out, all of which indicated the flatness of the earth and none of which were – to euphemize – easily reproduced by the few others who made the attempt. Many bogus scientists and charlatans before and since have exploited the way that, when we're told of an experimental result, we assume we're being told the truth; even if we could in theory perform the experiment ourselves to check it, in practice we don't.* As an example, Parallax claimed he'd shot cannonballs directly upwards and that – unlike the expectation for a rotating

---

* The same is true even of experiments reported in the scholarly journals, a fact that, *pace* the apologists for the scientific establishment's

spherical earth – they'd uniformly landed close by the gun
that had fired them, on occasion even lodging themselves
back in the barrel.

He lectured the length and breadth of England on
zetetic (i.e., geocentric flat-earth) astronomy; those lectures
were well attended and, by all accounts, extremely persuasive. The great popularizer of astronomy Richard Proctor
(1837–1888) conceded that, at least amid the immediacy of
the lecture hall, Parallax argued a very good case.

Of all the experiments done in an attempt to settle the
argument between planism and globism, the most famous
were those carried out on the Old Bedford Level, a roughly
10km stretch of canal in Cambridgeshire that ran dead
straight between two bridges (Welney Bridge and Old
Bedford Bridge). It seemed the perfect site for proving the
matter one way or another: if the earth were indeed spherical, the curvature should be easily detectable over a
distance of 10km. From 1838 onwards Parallax carried out
(or claimed to carry out) copious experiments on this
stretch of canal, inevitably demonstrating the fallacy of the
globular hypothesis: looking along the surface of the water,
he and his supporters were able to see objects that should
have been concealed by the curvature of the earth.†

Parallax's work at the Old Bedford Level might have
been happily ignored by the scientific establishment had it
not been for the actions of one of Parallax's more
hotheaded disciples, John Hampden (1819–1891).
Hampden was so convinced of the earth's flatness that in
1870 he made a public wager of £500 – then a princely sum
– to anyone who could come to the Old Bedford Level and

ability to regulate itself, has led to some very considerable chicanery.
See, for example, *False Prophets: Fraud and Error in Science and Medicine*
(1982) by Alexander Kohn, *Betrayers of the Truth: Fraud and Deceit in the
Halls of Science* (1982) by William Broad and Nicholas Wade, *The Great
Betrayal: Fraud in Science* (2004) by Horace Freeland Judson, and my
own *Corrupted Science* (2007).

† On occasion such observations need not have been fraudulent or self-
deluding. Atmospheric refraction over bodies of water can lead to a
curving of light rays such that the observer sees a bit, and sometimes a
lot, farther than would otherwise be the case.

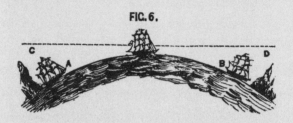

Above: Parallax demonstrates to his own satisfaction the implausibility of a globular world

Below: The rationale, according to Parallax, behind the first Old Bedford Level experiment

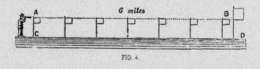

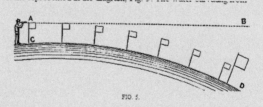

Above: Alfred Russel Wallace

Right: A version of Cosmas's world map (see page 30) produced by arch bogus theorist Ignatius Donnelly (see page 112)

demonstrate the earth's curvature. As it happened, Alfred Russel Wallace (1823–1913), co-originator with Charles Darwin (1809–1882) of the theory of evolution by natural selection, was severely broke at the time. Although initially reluctant to take candy from this particular baby, he rationalized to himself – after consultation with, among others, the "Father of Geology", Sir Charles Lyell (1797–1875) – that he might be able to win for science a war worth the winning while at the same time digging himself out of his financial hole. Hampden and Wallace agreed that J.H. Walsh (aka "Stonehenge"), editor of the magazine *The Field*, should be stakeholder. They also each appointed one referee: Walsh acted as Wallace's while Hampden named the printer, author and assiduous flat-earther William Carpenter (1830–1896) to act as his. (In the event Walsh could not stay for the entire session of experiments; his place was taken by local physician Martin Coulcher.)

The experiment Wallace had devised involved placing poles at regular intervals along the canal between the two bridges, each pole bearing a marker disc at a uniform height. If the earth were flat, the discs should appear all in a straight line to someone looking along the canal through a telescope; otherwise the curvature of the globe should, when the telescope was pointed at the farthest marker, make the intervening ones appear higher. While this proposal had seemed good on paper, in practice it was infernally difficult, squinting through the telescope, to work out which marker disc was which, and the experimenters agreed the result was inconclusive. Wallace then devised a simpler experiment and borrowed a better telescope, and a couple of days later they tried again, this time using just equal-height markers on the two bridges and on a solitary pole midway between them.

The observation conclusively demonstrated the curvature of the earth: exactly as predicted, the central marker appeared above the straight line between the markers on the bridges. Unfortunately, Carpenter inferred exactly the opposite from the result. The telescope that Wallace had borrowed included a cross hair that was quite irrelevant to the experiment. Looking through the eyepiece, Carpenter saw that the distance between the cross-hair and the image of the central marker was the same as that between the central marker and the one on the distant bridge, and claimed this as exactly the sort of perspective effect one would expect on the flat earth. Carpenter refused to be persuaded out of this, and eventually Hampden and Wallace resorted to the appointment of an umpire – in the event, the same J.H. Walsh who'd initially been Wallace's referee. It's very evident Walsh made every effort to be scrupulously fair in his judgement, even consulting a company of instrument makers to adjudicate in the matter of the cross-hair; he also published the reports of both Carpenter and Coulcher in *The Field*. Inevitably, Walsh finally found in Wallace's favour, and declared the £1000 to be his.*

It was at this moment that the future course of Wallace's life was devastated. Hampden, backed by Parallax (who'd played no part in earlier proceedings), claimed that Wallace and his "co-conspirators" were liars and cheats who were attempting to pull a curtain of false science over results that had in fact proved the planist theory. Hampden went to court; in 1876 the Queen's Bench decided, amid a haze of legal sophistry, that wagers on the earth's flatness or rotundity had no status in the law, and thus, since Hampden disputed the umpire's judgement, he could have his stake back.

So much for the money-making aspect of Wallace's participation. He assumed, however, that this was the end of it and that he could put several years of unpleasantness, dispute and legal expense behind him. He assumed wrong. Until the end of Hampden's life Wallace was subjected to a campaign of vilification, even including death threats. Several times Wallace was forced, for the security of his family, to have Hampden arrested, and both he and Walsh, who came in for his share of the vitriol, were driven separately to sue the man for libel. Here are some extracts from Wallace's account of it all in his autobiography *My Life* (1905):

---

* The cross-hair furore was long-lived. Much later, in 1897, the flat-earther James Naylor published an article claiming that the cross-hair had distorted the telescope's optics such as to create the illusion that the central marker was above the line of sight between the two bridges.

I will now briefly state what were Hampden's proceedings for
the next fifteen or sixteen years. He first began abusing Mr.
Walsh in letters, postcards, leaflets, and pamphlets, as a liar,
thief, and swindler. Then he began upon me with even more
virulence, writing to the presidents and secretaries of all the
societies to which I belonged, and to any of my friends whose
addresses he could obtain. One of his favourite statements in
these letters was, "Do you know that Mr. A.R. Wallace is allowing himself to be posted all over England as a cheat and a
swindler?" But he soon took more violent measures, and sent
the following letter to my wife:

"MRS. WALLACE,
    "Madam – If your infernal thief of a husband is brought
home some day on a hurdle, with every bone in his head
smashed to pulp, you will know the reason. Do you tell him
from me he is a lying infernal thief, and as sure as his name
is Wallace he never dies in his bed. –
    "You must be a miserable wretch to be obliged to live with
a convicted felon. Do not think or let him think I have done
with him.
                                        "JOHN HAMPDEN."

For this I brought him up before a police magistrate, and he
was bound over to keep the peace for three months, suffering
a week's imprisonment before he could find the necessary
sureties. But as soon as the three months were up, he began
again with more abuse than ever, distributing tracts and writing to small local papers all over England. . . .
    On January 13, 1893, he was brought up again for fresh
libels, and was again respited on publishing a fuller apology
and complete recantation of all his charges, as follows:-

"PUBLIC APOLOGY. – I, the undersigned John Hampden,
do hereby absolutely *withdraw* all libellous statements
published by me, which have reflected on the character of
Mr. Alfred Russel Wallace, and apologize for having
published them; and I promise that I will not repeat the
offence. – JOHN HAMPDEN."

This was published in several of the London daily papers and in various country papers in which any of his letters had appeared, and the judge gave him a serious warning that if brought up again he would be imprisoned.

Some months afterwards, however, he began again with equally foul libels, and I had him brought up under his recognizances, when he was sentenced to two months' imprisonment in Newgate.

But within a year he began again as violently as ever, and on March 6, 1875, he was indicted at Chelmsford Assizes for fresh libels, and on proof of his previous convictions and apologies, he was sentenced to one year's imprisonment and to keep the peace, under heavy recognizances and sureties, for two years more. . . .

Through the interest of his friends, however, he was liberated in about six months; and thereupon, in January, 1876, he brought an action against Mr. Walsh to recover his deposit of £500, and this action he won, on the grounds already stated; and as I had signed an indemnity to Mr. Walsh, I had to pay back the money, and also pay all the costs of the action, about £200 more. But as I had a judgment for £687 damages and costs in my libel suit against Hampden, I transferred this claim to Mr. Walsh as a set-off against the amount due by him. Hampden, however, had already made himself a bankrupt to prevent this claim being enforced, and had assigned all his actual or future assets to his son-in-law. . . .

. . . Hampden was by no means silenced. The very day after his recognizances expired, in 1878, he began again with his abusive postcards, circulars, and other forms of libel. In 1885 he wrote and printed a long letter to [T.H.] Huxley, as President of the Royal Society, chiefly on his biblical discussion with Mr. Gladstone, in a postscript to which he writes as follows:

"I have thoroughly exposed that degraded blackleg, Alfred Russel Wallace, as I would every one who publicly identifies himself with such grossly false science, which he had the audacity to claim to be true! If this man's experiment on the Bedford canal was founded on fact, then the whole of the Scriptures are false, from the first verse to the last. But your whole system is based upon falsehood and fraud, and refusal of all discussion; and such characters as Wallace seem to be your only champions." . . .

About this time he printed one thousand copies of a two-page leaflet, and sent them to almost every one in my neighbourhood whose address he could obtain, including most of the masters of Charterhouse School, and the residents as well as the tradesmen of Godalming. It was full of – "scientific villainy and roguery," – " cheat, swindler, and impostor." – "My specific charge against Mr. A.R. Wallace is that he obtained possession of a cheque for £1,000 by fraud and falsehood of a party who had no authority to dispose of it" – "As Mr. Wallace seems wholly devoid of any sense of honour of his own, I shall most readily submit the whole matter to any two or more disinterested parties, and adhere most absolutely and finally to their decision." – "I will compel him to acknowledge that the curvature of water which he and his dupes pretend was proved on the Bedford Level, *does not exist*! And this Mr. Wallace saw with his own eyes." And so on in various forms of repetition and abuse. . . .

One day about this time we happened to have several friends with us, and as we were at luncheon, I was called to see a gentleman at the door. I went, and there was Hampden! I was so taken aback that my only idea was to get rid of him as soon as possible, but I afterwards much regretted that I did not ask him in, give him luncheon, and introduce him as the man who devoted his life to converting the world into the belief that the earth was flat. We should at least have had some amusement; and to let him say what he had to say to a lot of intelligent people might have done him good. But such "happy thoughts" come too late. He had come really to see where I lived, and as our cottage and garden at Godalming, though quite small, were very pretty, he was able to say afterwards that I (the thief, etc.) was living in luxury, while he, the martyr to true science, was in poverty. . . .

The two law suits, the four prosecutions for libel, the payments and costs of the settlement, amounted to considerably more than the £500 I received from Hampden, besides which I bore all the costs of the week's experiments, and between fifteen and twenty years of continued persecution – a tolerably severe punishment for what I did not at the time recognize as an ethical lapse.

William Carpenter (see page 36), who seems by all accounts to have been a nice enough fellow, emigrated *c*1880 to the US, where he continued his activities as a flat-earth campaigner – although making his living primarily through his old profession of printer and also as a teacher of shorthand. In both countries he emitted a profusion of tracts and booklets. Probably his best-known work is *One Hundred Proofs the Earth is not a Globe* (1885), which went through several editions. Here are some of the "proofs":

1. The aeronaut can see for himself that Earth is a Plane. The appearance presented to him, even at the highest elevation he has ever attained, is that of a concave surface – this being exactly what is to be expected of a surface that is truly level, since it is the nature of level surfaces to appear to rise to a level with the eye of the observer. This is ocular demonstration and proof that Earth is not a globe.

17. Human beings require a surface on which to live that, in its general character, shall be LEVEL; and since the Omniscient Creator must have been perfectly acquainted with the requirements of His creatures, it follows that, being an All-wise Creator, He has met them thoroughly. This is a theological proof that the Earth is not a globe.

33. If the Earth were a globe, people – except those on the top – would, certainly, have to be "fastened" to its surface by some means or other, whether by the "attraction" of astronomers or by some other undiscovered and undiscoverable process! But, as we know that we simply walk on its surface without any other aid than that which is necessary for locomotion on a plane, it follows that we have, herein, a conclusive proof that Earth is not a globe.

63. It is a fact not so well known as it ought to be that when a ship, in sailing away from us, has reached the point at which her hull is lost to our unaided vision, a good telescope will restore to our view this portion of the vessel. Now, since telescopes are not made to enable people to see through a "hill of water," it is clear that the hulls of ships are not behind a hill of water when they can be seen through a telescope though lost to our unaided vision. This is a proof that Earth is not a globe.

75. Considerably more than a million Earths would be required to make up a body like the Sun, the astronomers tell us: and more than 53,000 suns would be wanted to equal the cubic contents of the star Vega. And Vega is a "small star"! And there are countless millions of these stars! And it takes 30,000,000 years for the light of some of those stars to reach us at 12,000,000 miles in a minute! And, says Mr. Proctor, "I think a moderate estimate of the age of the Earth would be

500,000,000 years!" "Its weight," says the same individual, "is
6,000,000,000,000,000,000,000,000 tons!" Now, since no human
being is able to comprehend these things, the giving of them
to the world is an insult – an outrage. And though they have
all risen from the one assumption that Earth is a planet,
instead of upholding the assumption, they drag it down by the
weight of their own absurdity, and leave it lying in the dust – a
proof that Earth is not a globe.

And so muddleheadedly on.

Back in the UK, Parallax had died in 1884. Not long
before his death he had founded the UK branch of the
Zetetic Society, in 1883; the US branch had been founded a
decade earlier, in 1872 in New York, by one George Davey,
with Parallax as its president. This was an era in which even
the prominent might be persuaded by the planist cause:
one such was the Transvaal President Paul Kruger
(1825–1904), although in 1900 he recanted when the skip-
per of a Dutch man-o'-war explained to him that maritime
navigation depended on the earth's sphericity.

The leadership of the Zetetic Society's UK branch was
eventually inherited, in 1893, by a dynamic and formidable
individual, Lady Elizabeth Anne Blount (dc1924), who
renamed it the Universal Zetetic Society. Describing herself
modestly as a geographer, explorer, mathematician, author,
composer and poet, she edited and published (and largely
wrote, often under the pseudonym Zeteo) the society's journal, Earth – Not a Globe – Review, alongside a flood of
pamphlets and magazines. Also from her prolific pen came
one of the 19th century's oddest novels, Adrian Galilio
(1898), a fantasy whose heroine is the ravishingly beautiful
Lady Alma. Like the novel's author, the Protestant Lady
Alma marries a much older aristocrat who's a Roman
Catholic; because he forbids her to mix with other
Protestants, she takes on a priest as a lover. The two para-
mours are shot by the priest's passionate housekeeper. On
recovery from their wounds, the two go their separate ways,
the priest living in Paris as a promiscuous roue and Lady
Alma adopting the persona of "Madame Bianka", in which
guise she carves out a successful career giving lectures on,
and singing hymns in praise of, zetetic astronomy.

Later Lady Blount and William Thomas Wiseman adapted the novel for the stage as the operetta *Astrea, or The Witness of the Stars*. This was not the first collaboration between the two. Lady Blount had the habit of filling the pages of *Earth – Not a Globe – Review* with her verses, and Wiseman had matched one of these epics, "The Nebular Hypothesis", to his own "Earth Not a Globe Waltz". (Even before that, Lady Blount had herself set the poem to music.) As one might guess from the poem's title, it forms an attack upon the then-accepted hypothesis for the solar system's origin, its condensation out of a cloud of gas and dust. Other poems by Lady Blount (and her fellow contributors to the *Review*; they included William Carpenter, who for the last few years of his life acted as US liaison for the Universal Zetetic Society) targeted evolution and the geological timescale, both of which were unacceptable to her because contrary to the Bible.

Unlike Parallax, who attempted to beguile the opposition with seeming science, Lady Blount was unashamed to admit she derived her planist views solely from her Christian belief in the literal truth of the Bible. This led her to be, as well as a planist, a young-earth Creationist. In the *Atlanta Constitution* she was reported in 1905 as saying:

> If the Bible is the word of God it is absolutely true. We must accept it as a whole or else accept none of it. We cannot divorce the religion of the Bible from the science of the Bible, hence the globists cannot be Christians – nor can Bible Christians be followers of [Sir Isaac] Newton's philosophy.*

Nevertheless, Lady Blount did offer supposed scientific arguments in favour of the flat earth – or, rather, counter-arguments against the evidences of the earth's sphericity. Some of these leave this reader, at least, scratching his head; the evidence of Foucault's pendulum, for example, she

---

* As cited by Christine Garwood in her fine book *Flat Earth* (2007), which tells of planist movements over the centuries and is invaluable in the cases of Lady Blount and Samuel Shenton. Not that Garwood is herself without blemish: at one point she claims, while discussing weightlessness in space, that "gravity does not exist far beyond the earth".

dismissed as worthless because it indicated only that the
pendulum moved, not that the earth did.* That ships disap-
peared from the waterline up as they vanished over the
horizon she claimed (as had Parallax before her) was merely
a perspective effect. And she went back to the Old Bedford
Level, where, rather than repeat Alfred Russel Wallace's
experiment, she devised one of her own: from one of the
bridges she hung a sheet that almost touched the surface of
the water, while a photographer named E. Clifton took a
picture along the waterline from the other bridge. In theory,
because of the earth's curvature, the sheet should have been
invisible from Clifton's position, but in fact his photograph
was claimed to show the sheet in its entirety. The obvious
explanation, despite accounts claiming that Clifton was a
globist and as mystified as anyone else, is that he cheated;
but this may be to malign him, in which case the photograph affords a genuine mystery – albeit not a mystery that
anyone's losing too much sleep over. An alternative explanation was implied in Robert Schadewald's description of
the photograph in *Worlds of Their Own* (2008): "I can't even
find the *bridge.*"

At roughly the same time that Lady Blount was crusad-
ing for planism in the UK, on the other side of the Atlantic
a religious zealot called Wilbur Glenn Voliva (1870–1942)
was running his planist community at Zion, Illinois. In 1895
the evangelical Christian and faith-healer John Alexander
Dowie (1847–1907) had founded the Christian Apostolic
Church and in 1901 the community of Zion. A decade later
Voliva, who had joined the Church in 1899, usurped

---

* Due to the rotation of our planet, the line of oscillation of a
pendulum left swinging undisturbed moves through 360° during each
24 hours. Carpenter, too, was sceptical about this evidence of the earth's
rotation, commenting in his *One Hundred Proofs*:

> 73. Astronomers [say] the table moved round under the pendulum,
> instead of the pendulum shifting and oscillating in different direc-
> tions over the table! But, since it has been found that, as often as not,
> the pendulum went round the wrong way for the "rotation" theory,
> chagrin has taken the place of exultation, and we have a proof of the
> failure of astronomers in their efforts to substantiate their theory,
> and, therefore, a proof that Earth is not a globe.

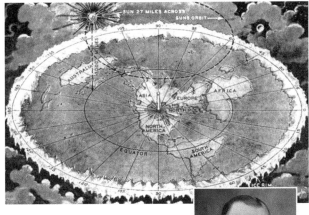

Right: Wilbur Glenn Voliva

Above: Voliva's cosmology depicted in the October 1931 issue of *Modern Mechanics & Invention*

Dowie's leadership – although "usurped" is perhaps too harsh a word in the circumstances: on Dowie's suffering a stroke, Voliva was called back from a mission he'd been undertaking in Australia; arriving in Zion, he found Dowie incapable because of the stroke, and so not unreasonably he took over supreme power. It was shortly after this that the schools in Zion began to teach that the earth is flat.

Under Voliva's rule Zion for the most part prospered – despite strict laws governing matters the rest of us might regard as trivia: such offences as drinking, smoking, swearing or (on a Sunday) whistling brought heavy fines, and women on horseback were permitted to ride sidesaddle only. The prosperity arose largely because of the Zion Fig Bar, an item of confectionery that was hugely popular far beyond the city limits.

Voliva travelled several times around the world trying to persuade other people to accept the flatness of the earth.

He also believed the world's end was nigh, that soon the
planet would be consumed by an inferno. He told his
followers this sorry destruction would occur in 1923, then
(when 1923 proved an Armageddon-free zone) in 1927,
then in 1930, then in 1935; between 1935 and his own
demise in 1942 he had the sense to keep quiet about his
premonitions.

In the early 1920s Voliva started the radio station
WCBD through which he broadcast 5000W of derision and
defiance directed towards what he called The Devil's
Triplets: scientific astronomy/cosmology, evolution and – a
slightly odd addition – higher criticism. Within the city his
power, though occasionally challenged in minor matters,
was for the most part virtually unlimited. Yet Robert
Schadewald (in *Worlds of Their Own*) pointed out something
curious:

> Though [Zion] was for more than thirty years the flat-earth
> capital of the world, Voliva and his colleagues produced
> remarkably little flat-earth literature and contributed
> absolutely nothing to flat-earth theory. The May 10 1930 issue
> of [the community's journal] *Leaves of Healing* was devoted
> entirely to the earth's shape, and the church also reprinted
> Carpenter's *One Hundred Proofs that the Earth Is Not a Globe*. But
> that was all.

Voliva had laid out his general principles to the population
of Zion in December 1915:

> I believe this earth is a stationary plane; that it rests upon
> water; and that there is no such thing as the Earth moving, no
> such thing as the Earth's axis or the Earth's orbit. It is a lot of
> silly rot, born in the egotistical brains of infidels. Neither do I
> believe there is any such thing as the law of gravitation. I
> believe that is a lot of rot, too. There is no such thing! I get my
> astronomy from the Bible.

The edition of *Leaves of Healing* to which Schadewald
referred focused on Voliva's interpretation of the Bible's
statements on astronomy – in particular on this supposedly
newfangled notion that the earth was spherical, not flat. He

spelled out his four proofs that the earth could not be a sphere spinning on its axis, as the astronomers claimed:

❖ The wind would always blow in the same direction, opposite to the direction of the earth's spin.

❖ If we jumped up in the air, remaining aloft for only a fraction of a second, we'd land far from where we started. Diving from a board would be suicidally dangerous, because one would inevitably land outside the swimming pool.

❖ We'd find it difficult to walk or otherwise travel contrary to the direction of the earth's spin.

❖ Objects on the surface of the earth – ourselves included – would soon be thrown off into the depths of space by centrifugal force.

Since none of these phenomena are observed, the earth must be stationary.

Voliva also made much of the fact that past greats in astronomy's history had "disagreed" over some fairly fundamental details; for example, Nicolas Copernicus (1473–1543) had thought the sun to be stationary while Sir William Herschel (1738–1822) believed the solar system as a whole to be moving through space. A rational observer might remark that Herschel was born nearly two centuries after Copernicus died, and that astronomy had advanced a little in the interim, but not Voliva:

> It is asserted by these advocates of the Copernican system of astronomy that it is an EXACT SCIENCE – and yet these two great men, Copernicus and Herschel, contradict each other, Copernicus saying that Herschel is a liar and Herschel saying that Copernicus is a liar – and Voliva agrees that they are both right!

Of course, Voliva did not confine his teachings to planism. In 1921 he published *Handbook and Guide to Hell*, advance information on which was given in a *New York Times* report:

VOLIVA PICTURES HELL.

Tells His Followers Topers Will
Stew In Their Favorite Juice.

ZION, Ill., Jan. 15.—Residents of Zion received new light on the terrors of the infernal regions today, when Overseer Wilbur Glenn Voliva issued advance sheets on a "Handbook and Guide to Hell," based on what he termed "helligrams," which he said he had received recently.

"Every sinner is going to be punished with an overdose of his own sin," Voliva declared. "A tobacco smoker will be locked up in a den full of tobacco smoke. A chewer of the filthy weed will be immersed to his neck in a vat of tobacco juice. A drinker will pass his term of purification in a natatorium filled with beer, wine and whisky."

Perhaps wisely, Voliva did not outline the "overdose" punishment awaiting the fornicator.

After Voliva's death, flat-earth beliefs seem to have died out in Zion fairly quickly. Approximately contemporary with Voliva but in the UK was the Somerset builder and inventor William Edgell (d1935) who, initially under the pseudonym William Westfield, self-published in 1914 (with revised editions in 1919 and 1927*) the chapbook *Does the Earth Rotate? NO!* His aim, apparently inspired by a promise he made to his dying planist father, was to persuade the UK education authorities to abandon the hypothesis of the earth's motion, both its rotation and its orbit around the sun; and he offered numerous simple experiments to demonstrate his point.† For example:

Here is another positive proof that the earth cannot rotate. In the Desert of Sahara, the length from east to west is 3,000

---

* The title page of the 1927 edition claims that he was the inventor of "the free-wheel for bicycles, the automatic weighing machine, the coin feed machine, the airless tyre, &c."

† I haven't been able to lay hands on a copy of Edgell's book myself, and so am indebted to the brief analysis posted by Alfred Armstrong on his Odd Books blog: http://oddbooks.co.uk.

miles, its average breadth 900 miles, and its area 2,000,000 square miles. Rain falls on this desert at intervals only of five to ten or twenty years. If the earth rotates over 10,000,000 miles* daily, and in addition makes another movement round the orbit and sun yearly, how can this large desert escape the rain from the heavens for years at a stretch, while other places receive the rain regularly?

Edgell seems, here, to have taken it as axiomatic that the heavens were fixed and that rain fell at different times from different parts of them. In such a model, clearly any particular rainfall would be experienced over a large area of the earth's surface as the earth spun beneath. The fact that this didn't occur was Edgell's proof that the earth did not in fact spin.

Another proof involved observing the pole star, Polaris, through a long tube mounted in his garden. By looking through this tube at the small area of the night sky visible at its far end, he satisfied himself that Polaris – which he demonstrated elsewhere to be a mere 8000km distant – did not move as seen from earth. This implied that both Polaris and the earth were fixed in space.

Of course, Edgell's stationary earth was also flat. Like other planists before and since, he was left with the problem of explaining the sun's rising and setting, and its seasonal behaviour. For the former phenomenon, it was all very well for planists to explain that the sun moved backwards and forwards above the earth's disc, but this begged the question as to why observers in the UK could not see low in the sky, at their midnight, the distant sun as it hovered directly over New Zealand, giving that country its noon. Edgell pointed out that, although the earth might be flat, its surface was corrugated by mountains and other obstacles, including, since the advent of human civilization, cities. As the sun shone down on lucky New Zealand, it would be at such a low angle in the sky from the UK that *something in the*

---

* Edgell later "corrected" this figure to 1,555,200 miles. Of course, the *true* figure equals the earth's equatorial circumference, which is about 40,000km (25,000 miles) – very substantially less than even Edgell's reduced figure!

intervening landscape would be bound to get in the way. If mountains and cities wouldn't oblige, then the obstacle might be the horizon, for "[a]ll readers are aware that mountains and hills and horizon are common in all countries". Presumably that damned intrusive horizon was responsible for mariners' claims to have seen the sun set at sea.

The conclusion of Edgell's book should be kept from the eyes of antiscientific cost-cutters like Senator James Inhofe (b1934), who might latch onto its proposal:

> With all due respect to astronomers' prophecies of future happenings as to comets, readers will see their judgement as to distances and earth rotation cannot be relied upon. May I ask, is it worth while keeping a large staff at our Observatories, or anyone working at a false and unreasonable theory . . .?
>
> A considerable sum of money can now be saved by greatly reducing the staff at Observatories in this country, and undoubtedly the Government will be convinced that the proofs given in this small book is [*sic*] overwhelming against the enormous distances given by astronomers, and that the earth rotation theory is absolutely disproved.

That flat-earth ideas still have any adherents at all is due to the determination of a few diehards, notably Samuel Shenton (1903–1971), aided by his wife Lillian, in the UK, and Charles Johnson (1924–2001), aided by his wife Marjory (d1996), in the US. (There was also, from the 1970s, the Canadian Flat Earth Society, but this organization's *raison d'être* was the philosophical one of rocking complacency's boat rather than any expression of genuine cosmological belief.)

Samuel Shenton was a signwriter in Dover, UK. He came to planism rather late in life, and via an odd route. He was interested in developing a better cargo-carrying aircraft than the existing airships and aeroplanes. His notion was that all an aircraft really needed to be able to do was to rise to altitude and hover there, while the earth spun beneath it, until the time came to descend to the destination. Of course, this would work only if departure and destination were at the same latitude, but even so an aircraft

constructed on this principle and according to Shenton's
design would represent a huge advance in air transportation – just think of the fuel savings alone! Strangely, none of
the authorities to whom Shenton sent his blueprints were as
ecstatic and grateful as he'd anticipated, and he began to
suspect Establishment science had something to hide. He
discovered his idea had been had before, by one of Lady
Blount's supporters, and this led him to study everything he
could find on zetetic astronomy. The solution to what it was
Establishment science had been covering up became
obvious: the reason they weren't keen on his aircraft was
that its functioning depended on a rotating earth. Should it
ever be built, its non-functioning would be dramatic
demonstration of the earth's non-rotating flatness. No
wonder the bureaucrats had put such obstacles in the way of
the design's development!

In certain obvious respects Shenton's cosmology resem-
bled those put forward by his planist predecessors: The
north pole was at the centre of the terrestrial disc and the
south pole was a great wall of ice that ran all the way round
the disc's circumference. Sun, moon and stars were small
and relatively nearby – the sun, for example, was some
4800km distant and about 50km in diameter. The moon,
which shone with its own light rather than merely reflecting
the sun's, was the same size but even closer, at a distance of
4100km. Before him, Edgell had reckoned the sun to be
rather closer, some 4000km away.

It isn't only planists who've thought the moon, sun and
stars are far closer to us than astronomy suggests. George
Bernard Shaw (1856–1950) believed the moon was only
about 50km distant, implying it could be only about 600m
across. Charles Hoy Fort reckoned the moon must be about
160km across and thus some 18,500km distant (see pages
97-8 for a fuller explanation). Much later, writing in 1964,
the ancient-astronaut theorist W. Raymond Drake agreed
with Fort's figure for the lunar diameter but not with that
for the distance; instead Drake maintained the moon
appears a lot bigger to us than it actually is because the
earth's atmosphere acts as a giant lens, magnifying the lunar
image by a factor of as much as 20. What seems particularly

bizarre about Drake's claim is that by 1964 several astronauts had left the earth's atmosphere yet none had noticed
the moon abruptly appearing smaller.

According to Shenton, the flat earth and its companion
bodies lie motionless at the bottom of one of countless pits
that pock the surface of the infinite plane that is the
universe. There the earth floats on water, some of which
seeps through the disc to form rivers and seas.

Shenton introduced an intriguing new explanation for
the apparent daily and annual motions of the sun. The sun,
which stays forever above the plane of the earth, emits light
in a beam like a spotlight's. It orbits not around the earth
but (as does the moon) in horizontal circles of varying diameter above the earth. During what we would describe as the
northern-hemisphere winter, the circle the sun is describing
is a large one, so that the main benefits from its heat and
light are enjoyed in the region of the flat earth's outer
circumference – i.e., the wall of ice that forms the south
pole. Six months later the sun is performing a much tighter
circle, so the regions closer to the disc's centre benefit.

In 1956 Shenton founded in London the International
Flat Earth Research Society (IFERS), with one William Mills
as its president. Among those present at the first IFERS
meeting was the popular amateur astronomer and broad-
caster Patrick Moore (b1923), who later described the experience in his book *Can You Speak Venusian?* (1976), where he
also cited an IFERS pamphlet:

> The International Flat Earth Society has been established to
> prove *by sound reasoning and factual evidence* that the present
> accepted theory, that the Earth is a globe spinning on its axis
> every 24 hours and at the same time describing an orbit round
> the Sun at a Speed of 66,000 m.p.h., is contrary to all experi-
> ence and sound commonsense.
>
> In ancient times the Earth was regarded as plane, and
> this is expressed in all literature up to a few hundreds of years
> ago. The theory has fallen into disfavour, owing mainly to the
> dogmatism of modern science and popular education in
> schools, which leads to prejudice in favour of the globular
> theory from the start.
>
> It is always a pity to allow false theories to pass unchal-

lenged, and it is hoped that the Flat Earth Society will do much to undo the harm that has been caused.

Remember that the truth of the plane figure of the Earth can be shown by *irrefutable evidence*, and anyone who is interested in becoming a member is asked to contact the President or the Organising Secretary. In future, it is hoped to hold regular meetings of the Society.

Less than a year after the founding of IFERS, *Sputnik 1* went into orbit – or not, according to Shenton. Instead, like the sun, it went round and round in circles above the plane of the earth. Just a couple of months after *Sputnik*'s launch, William Mills died, so the onus fell fully on Shenton to continue IFERS's evangelizing work, which he did willingly in clubs, pubs, classrooms, astronomical societies, churches and many other venues all over the UK off and on for the rest of his life.

An early concern was the dangers astronauts might encounter should they attempt to land on the moon; Shenton did not say that the moon was immaterial, but he did seem to think it had a sort of insubstantiality to it, maintaining that on occasion people had been able to look right through it to the stars beyond. And Shenton offered an explanation for the *Apollo 1* disaster of January 1967 that decades later would be eerily echoed by the likes of Jerry Falwell and Pat Robertson when ascribing blame for the disaster of September 11 2001. As Shenton explained in an (unpublished) 15 April 1967 letter to the editor of the *Daily Express*, cited in part (like the IFERS flyer above) in Garwood's *Flat Earth*:

Dear Sir,

Are we so fascinated by "science" that we fail to realize the true character of the contending forces around us? Our American brothers emphasize their motto "In God We Trust". But in their modern efforts to climb into the heavens they affront God by naming their thrusting ventures "APOLLO" the pagan mythical god who in ancient times was associated with sudden death! The tragic sudden deaths of the three American astronauts and today's equally tragic death of

Russia's representative[*] may give watching Christians time to realize that, as stated, God is not mocked.

Yours truly

S. Shenton.

On Shenton's death in 1971 his widow Lillian was left with the problem of what to do with all the IFERS records. In due course she decided they should be shipped to the USA, to Charles K. Johnson, who thereby assumed the leadership of the IFERS. It was a post he was to retain for the rest of his life. At one stage the quarterly magazine he published, *Flat Earth News*, had as many as 3500 subscribers.

Born in Texas in 1924, Johnson was converted to planism at the age of eight. Many years later he described the reason. "When I was at school the first maps I saw were flat. Then Roosevelt flooded all the classrooms with globes." Spinning the globe on display in his own classroom was enough to convince him that talk of a spherical, rotating world was nonsensical: it was patently obvious everything on the earth's surface would fly off, no matter the scientists' appeal to an ineffable force they called gravity. Just to be certain, he made observations of a nearby lake to see if he could detect the earth's curvature. As he'd expected, not a trace. "Obviously water's flat, isn't it? They're trying to tell you water's bent?" Much later in life he'd perform similar "tests" on Lake Tahoe and the Salton Sea, with similar confirmatory results.

His wife Marjory came from Australia and was fond of attesting that she knew the earth wasn't flat, otherwise she'd have lived her childhood upside-down. Her death in 1996 seems to have been hastened by the fire that took their home in 1995, which fire also destroyed all of the IFERS records and membership rolls. In due course Johnson was evicted from the trailer he and Marjory had moved into after the fire, and for the last few years of his life he lived with his brother and worked to bring the IFERS membership back up to about 100.

---

[*] The cosmonaut Vladimir Komarov, who died on April 24 when *Soyuz 1*'s parachutes failed on the craft's return to earth.

Clearly NASA's activities presented something of a challenge to the Johnsons, as to all other planists. Johnson himself was convinced the televised scenes of space explorations, including the various Apollo landings on the moon, were scripted by Arthur C. Clarke and filmed by Stanley Kubrick. (On being told of this, Clarke joked that he wanted his royalties.) The rights to this drama had been settled on the US rather than the USSR during a Kennedy–Kruschev summit in which it was also agreed that, in exchange, Cuba would remain Communist. In case this might sound like the kind of nonsense even flat-earthers would have difficulty swallowing, consider a 1994 *Washington Post* poll that discovered no fewer than 9% of the US public shared the view that the lunar landings had been faked.*

Internet sites such as http://theflatearthsociety.org/ forum keep the planist flag aloft, although of course it's difficult to ascertain how sincere the contributors to such websites might be: as we read their comments it's obvious that some participants are just there for the giggles, while other items, often far funnier, might or might not be seriously intended – for example:

> [T]here is no optical light, there is just light and theres [*sic*] no other type of light unless you start talkling [*sic*] about energy saving lightbulbs compared to other types of light bulbs.

Also among those still carrying the flag for planism in the wake of Johnson's 2001 death is John Davis, a Canadian-born computer scientist now working in Tennessee. He believes the flat earth extends infinitely in all directions, and is at least 9000km thick. This raises the notion that it's by no means impossible we'll one day have technology capable of drilling right through the flat earth and out the other side. Who could speculate what we might find? Might we

---

* A depressing number of people believe the plot of the 1978 movie *Capricorn One*, in which a Mars mission has to be faked, is perfectly plausible science fiction rather than escapist fantasy. Ironically, one of the astronauts supposedly killed on re-entry, and in reality now on the run for fear of being bumped off by feds eager to stop the truth coming out, was played by O.J. Simpson.

discover gravity works in the opposite direction there, so we could walk around on a second infinite plane, or does the direction of gravity's pull continue downwards? If so, where to?

In the UK James McIntyre is another planist stalwart. By contrast with Davis's ideas of an infinite earth, he believes the world is a disc of some 40,000km diameter with Antarctica a solid wall of ice extending all around the circumference. As for space photos showing a spherical earth, McIntyre comments: "The space agencies of the world are involved in an international conspiracy to dupe the public for vast profit" – that vast profit presumably arising because only a tiny fraction of all the billions supposedly spent on space research needs to be spent faking Martian landscapes and Hubble deep-range photos.

Some people really believe the earth is at the centre of the universe. This notion

is, of course, inherent within most hypotheses of the earth's flatness: that our world is stationary and all the celestial bodies revolve around it. Geocentrism was certainly the view of the Bible writers,* and as a consequence – despite the understanding by at least some of the ancient Greeks that the earth went around the sun – remained prevalent in the mainstream of European thought until, famously, the Copernican Revolution and the publicity given to the Copernican hypothesis by Galileo. The further work of Kepler and Newton would seem to have put the final nails in the coffin of geocentrism some centuries ago. However,

---

* "That the Bible is overtly geocentric has been noted by believer and unbeliever alike," states Gerardus D. Bouw on an early page of his *A Geocentricity Primer* (2004). Many evangelicals today, confronted by the geocentrism and planism evident in the Bible, are waging a rearguard action to claim this is not really what it says at all. It's to the credit of the planists and the geocentrists that, as it were, they put their cosmology where their mouth is. For more, see pages 28-30.

as is the case with planism, if you made that assumption you were wrong: in the early years of the 21st century about one American in five believes the sun goes round the earth, with a further 9% of Americans being unsure which goes round the other, if either.

If modern geocentrism has a major overlap among planists, its other main association is with Creationism. The vast majority of Creationists accept that the earth orbits the sun while the sun is in motion around the galactic centre, and so on, just as scientific astronomy indicates; however, it's probably safe to say that *all* geocentrists are also Creationists. Both schools of thought draw their inspiration from the dogma that the Bible – notably *Genesis* – is literally true, with the non-geocentrist Creationists being a bit more flexible about the meaning of the word "literally".* The unease of allowing only a strict interpretation of some biblical statements while being more open to accepting others as merely symbolic was pinpointed by the father of modern geocentrism himself, the Dutch–Canadian schoolteacher Walter van der Kamp (1913–1998), in the first of his three essays comprising *The Whys and Wherefores of Geocentrism* (1988-9):

> . . . in 1963 I eagerly joined Walter Lang's Bible Science Association and the Creation Research Society.† Still blissfully unaware of the affray in which I thus would entangle myself, I read everything about the "how-to-understand-*Genesis*" question that I could lay my hands on. But although I shared the creationists' high view of Scripture and applauded their zeal, I slowly became aware that there was an inconsistency in their treatment of the first chapter of *Genesis*.
>
> The creationists stoutly denounced worldly biological input with regard to the verses 11–13 and 20–31 of *Genesis* 1, but apparently no scruples impeded them from more or less

---

* It's quite probable there are geocentrists who derive their views from other than religious sources, but they're hard to locate. Also, while there are geocentrists whose ideas are rooted in Judaism, Islam and Roman Catholicism, the vast majority of known modern geocentrists in the West are American Protestants.

† Both stalwart institutions of "creation science" – JG.

reconciling the pronouncements of modern cosmogony and
cosmology with the matter-of-fact statements in *Genesis* 1:1–9
and 14–19. Yet, it dawned on me, when those theories are
compared next to the plain text of Holy Writ, we cannot fail to
see that they contradict God's Word as brazenly in the matter
of his preparation of the Earth as post-Darwinian biology did
with regard to the emergence of life on that same Earth.

Modern geocentrists embrace various different models for
the workings of the universe. Rather like their counterparts
in the Creationist movement, they talk much about science
but rarely focus on it, instead reverting ever to their read-
ing(s) of the Bible, which interpretations they insist can be
backed up by the discoveries of science. Pressed to identify
the discoveries to which they're referring, their tendency is
to ignore the great bulk of scientific knowledge in order to
nitpick over difficulties of detail they perceive science to
have. Thus there's much shouting about the "fact" that
science has never offered any proof of the earth's motion
and, further, Relativity's statement that, since no frame of
reference within the universe is preferable to any other, all
motion is relative – that it's as meaningful, mathematically,
to say the earth is at rest while everything else moves as it is
to say that the earth moves in relation to anything else.

The system of geocentricity most significant today is
that developed by the great Danish astronomer Tycho
Brahe (1546–1601), who was Kepler's employer and mentor
before Kepler embraced the newfangled Copernicanism
and brilliantly refined it.* Tycho was unwilling to accept
that the earth moved; at the same time he realized that
existing Ptolemaic models – where everything went round a
stationary earth, the stars in circles and the planets in circles
that had to be modified by further ancillary circles (epicy-
cles), and so forth – were so cumbersome, and demanded so

---

* Copernicus thought the planets, earth included, followed circular
paths around the sun. This misconception led very quickly to disparities
between astronomical observations and the predictions of theory. It was
Kepler who realized the planetary paths must be ellipses. Not long
afterwards, Newton worked out laws of motion from which Kepler's
empirical workings, compiled over decades, could have been derived
promptly. Such is the cruelty of science's advance.

many special considerations if their explanations were to hold water, that they couldn't be a true reflection of reality. The system he proposed was a sort of halfway stage between the old geocentrism and the new heliocentrism: while the planets went around the sun, as Copernicus claimed, the sun, with its circling retinue of planets and comets, went around the earth. In the much farther distance, the stars likewise went around the earth. Tycho spent much of his later life attempting to prove this model was the reality despite his (excellent) astronomical observations telling him otherwise.

Until relatively recently the Tychonian system was merely an intriguing little dead end of scientific history, but the work of Walter (originally Wolter) van der Kamp was to reanimate it. He arrived in Canada from the Netherlands in 1955 and, after becoming involved in Creationism, published in 1968 (with a draft circulated in 1967) the pamphlet *The Heart of the Matter*, advocating the geocentrist view. He later admitted that almost the only sales of this tract had been to "compassionate friends and acquaintances". Nevertheless, in 1971 he founded the Tychonian Society, an informal organization whose agenda was to press the case for Brahe's model of geocentrism through various means. Van der Kamp's religious motivation for geocentrism is everywhere evident; as a single example, later he would write about "the theorizing and marketing of the post-Copernican view" being "undertaken by the minions of mankind's Archenemy".

The most notable servant of the Society's agenda was *The Bulletin of the Tychonian Society*. This journal was initially merely "handwritten and crudely multiplied on a Gestetner", according to van der Kamp, with "issue number 5 proudly announcing a modest fifty-copies edition". He didn't charge recipients for copies of the *Bulletin*, but did request donations; his deal with himself was that, if donations stopped covering costs, he'd quit producing the journal – which indeed he did for a period between 1971 and 1974. Issue #6, dating from the latter year, while still stencilled introduced a technological innovation: it was typed. Since then the *Bulletin* has never looked back, although it changed its name in 1991 to *The Biblical Astronomer* when the Tychonian Society's name was likewise changed to The Association for Biblical Astronomy by van der Kamp's successor (from 1984), Gerardus Bouw (b*c*1945).

Bouw – who, bogglingly, has a BSc in astrophysics from the University of Rochester and a PhD in astronomy from Case Western Reserve University – went through a period of atheism before rediscovering God:*

> Tired of sin and disillusioned with man, in May of 1974 I happened upon a science fiction work by [Robert A.] Heinlein entitled *Time Enough for Love* [1973]. Before long I was frustrated in the reading: Heinlein was obviously trying to write a bible for our times but all his gems came from the Holy Bible. I never finished Heinlein as I decided to go directly to the source itself.
>
> So I started a critical reading of the Bible, from cover to cover, searching for inconsistencies and any contradictions between an infinite God and the God of the Bible. Fortunately, God was watching out for me in that the only Bible I owned was an Authorized Version, the only English Bible free of such contradictions. Any other version and I would have been left with no alternative but agnosticism.

He became a born-again Christian in January 1975, and set out to prove the Bible right in all things, and science wrong. The conversion was apparently not an instantaneous one:

> At the time, I attended a Free Methodist Church in Rochester, and that March (1975), the Sunday school superintendent presented a little ditty to the children entitled *I Didn't Come From A Monkey, No, No*. Now it so happened that I had been working on a theistic evolutionary model in which the major phases of the Big Bang and evolution all happened within one day, with eons between the days. In all modesty, it was the best theistic evolutionary model I've ever seen, bar none. And now this man was going to tell *me* that evolution was not true? I snickered to myself: "Of course we didn't come from monkeys, everyone knows that we came from apes," and I resolved to correct him privately after Sunday school. Well, he had some

---

* This and succeeding quotes from Bouw are taken from his own "Testimony" on the geocentricity.com website.

tracts, specifically, one by Duane Gish [b1921] entitled *Have You Been Brainwashed?* [1974], and I postponed correcting him until after reading it.

You know, I never did correct that Sunday school superintendent. I stood corrected instead. I abandoned my theistic evolutionary model for what it was: dead wrong. When I became an atheist it was because I recognized that evolution and the Bible won't mix. I'd forgotten that, and was trying to "correct" the Bible to fit evolution, and not the other way around. True, I was trying to keep the corrections to an absolute minimum, but even so, I was trying to correct that which was perfect. And so it was that I became a Special Creationist (meaning that the universe is no more than 6,000 years old).

Bouw's geocentrism differs from that of his mentor. Where van der Kamp subscribed to the Tychonian system, Bouw adheres yet more literally, as one might guess from his statements above, to the Judaeo-Christian scriptures. Well, sort of. *Genesis* states fairly unequivocally that God created the earth before the lights of the heavens, so that the firmament is really part of the earth's accessories, as it were – the lighting fixtures put into the shell of the newly built house. On this Bouw disagrees: he holds that the heavens came first, the earth later. From here, however, his views diverge radically from those of the scientific community.

According to Bouw, the universe is filled with a solid yet completely transparent medium called the aether.* In numerous concentric spherical shells of this material – rather like the crystal spheres of yore – the stars are embedded. Outside all the rest there's a final, likewise rotating, material shell, whose main role seems to be as a gravitational engine whereby lots of other physical phenomena can be explained. An example of these is the Coriolis Effect,

---

* Not to be confused with the luminiferous aether, the universe-filling stuff that for a while physicists postulated as the medium *via* which electromagnetic waves travelled through space. It's a different aether altogether.

Bouw later decided he'd prefer the rigid, non-luminiferous aether to be instead called the "firmament". This seems one terminological hazard too many, so I've stuck with "aether".

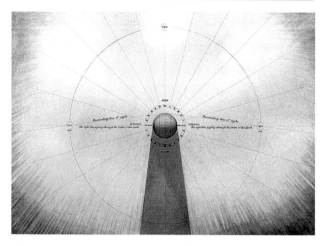

Geocentrism portrayed in Isaac Frost's *Two Systems of Astronomy* (1846)

whereby (in orthodox science) the earth's rotation is manifested in such phenomena as the differing preferred directions of rotation in hurricanes/cyclones in northern and
southern hemispheres.

One of the major difficulties of this model is that in real
life the stars are not fixed: they move about relative to each
other. Because of the great distances of the stars, in most
instances these movements (called proper motions) can be
detected only with highly sensitive equipment using observations spread out over a period of decades if not centuries.
But some proper motions – where the star's relatively
nearby and is travelling quickly – are much more easily
observed, and they show no signs that the star is having to
push its way through any kind of viscous medium. Such data
are obviously difficult to reconcile with the notion of a rigid
aether.

The proper-motion problem is not eased if one revisu-
alizes the size of the universe relentlessly downwards, as
some geocentrists have done in order to deal with another
severe difficulty the theory has. If the universe has anything
like the scale orthodox astronomers claim, it will take the
light from the stars in the outermost shells significantly

longer to reach us than the light from the stars in smaller, closer shells. Yet overall – ignoring for the moment individual proper motions, planetary motions, etc. – it can be seen that the heavens revolve about the earth as one. If the velocity of light were infinite the problem would be trivial: outer shells would move faster than inner ones in a regular fashion, so that the illusion was created of a unified rotation. But the velocity of light isn't infinite, and so the whole postulated situation becomes extraordinarily more complex. Even could one work out the math involved in getting the shells to rotate at relative speeds that would permit the illusion, one would be left at a loss to show how that system could continue to function in light of the observed variations in the length of our world's day – those tiny variations that orthodox science attributes to phenomena like earthquakes, violent weather systems, nutation, etc. How do the more distant shells know they have to briefly speed up or slow down in order that the stars embedded within them may retain their relative positions as observed from earth?

A further concern is that, of course, even if the inner concentric shells are fairly close to us, the outer ones, because of the sheer numbers of stars in the sky, must still be distant enough that they'd have to travel at many times the velocity of light in order to go round the earth once each day. And, anyway, the inner shells can't be that close because they're further away than the planets. The planets cause another problem: through radar and other means we know not just planetary distances but also the velocities with which the planets are moving through space, and those velocities are quite simply insufficient for the planets to make the circumterrestrial voyage every day.

Baffled by the intractability of such riddles, many geocentrists have turned to Relativity as a means of explaining how the earth can remain motionless while the rest of the universe dances around it. A tenet of Relativity is that there's no such thing as absolute motion: you can define an item's movement only in relation to other items. This means, trivially, you can assert that you yourself (or the planet you live on) remain motionless while everything else is moving – and of course the geocentrists have seized upon this notion as if it justifies their own tenet that the earth is stationary. (Just before you're tempted to waste thought on the latter argument, remember that mathematically the statement that your car remains still while the world moves is as meaningful as the reality.) Similar Relativistic reasoning leads to the conclusion that, since the universe has no centre-point, then every point in the universe has equal claim to being its centre. As a result, the geocentrists assert that Relativity supports the idea that . . . no need to continue!

Like their colleagues and frequent allies in the fields of Creationism and planism, the geocentrists are immune to any scientific demonstration that might refute their obsession. Just as the Creationist will complain about, say, the lack of transitional forms in the fossil record even as examples of transitional forms are placed in front of him, on the grounds that no two transitional forms can ever be *transitional enough*, so the geocentrist will either interpret contrary scientific evidence as in fact supporting geocentrism or just blithely ignore it. Van der Kamp summed it up in his *The Whys and Wherefores of Geocentrism*:

> I quested far and wide, and everywhere I came face to face with a dishonesty, a misleading practice, I never had thought possible. None, but none among the fanciful assertions of the believers in Galileo's sun-centered astronomical gospel has ever been proven. It is after Eve's seduction in the Garden of Eden the most monstrous deception ever foisted on mankind. The procedure for bringing it about has been an old, but effective one. The late non-lamented Joseph Goebbels, Hitler's henchman for propaganda, used it with great success in Nazi Germany: proclaim a lie again, and again, then in due course all people will believe you. . . . With the Great Lie of Galileo, lest perchance God be glorified again, Homo Sapiens *must* be kept brainwashed.

Not a single claim of modern scientific astronomy "has ever been proven"? Where does one begin to argue with such a profoundly impervious mindset?

In fact, there's a very simple demonstration that the earth is in motion. In the same way that you see more bugs flattened on the front window of your car than on the back, you'd expect to see more meteors when you were on the front side of a moving earth than when you were on its rear side. Wherever you happen to be, between your local midnight and the succeeding noon you're on the front side of the orbiting earth. Sure enough, statistically you'll see more meteors per hour after midnight than before. Such a disparity is inexplicable without special pleading unless the earth is in motion, both rotating on its axis and travelling through space.

In 1980 the geocentrist Richard G. Elmendorf offered through the pages of *The Bulletin of the Tychonian Society* a prize of $1000 for "Scientific Proof Positive that the Earth Moves". So far as I can establish, this prize has yet to be awarded. Perhaps modern scientists, wiser than Alfred Russel Wallace, realize the mare's nest into which any attempt to claim it might lead them.

Some people really believe we live on the inside of a hollow earth. Well, not so

many of them any more, not since 1961 when the last four members of the Koreshan community of Estero, Florida, deeded the community's property to the state, and anyway hollow-earthism has never been exactly a large-scale movement.*

In 1894 the amateur scientist, herbalist, quack and alchemist Cyrus Reed Teed (1839–1908), who came to be known as Koresh (the Hebrew version of Cyrus), led believers in his cosmological theory, Cellular Cosmogony, to set up the Estero community, which was intended to be a New Jerusalem. Also known as Koreshan Universology, the

* For further information on the more conventional hollow-earth theories, see my book *Discarded Science* (2006) or the discussion in Martin Gardner's *Fads & Fallacies in the Name of Science* (1952). There's also a very good discussion online by Donald E. Simanek: "Turning the Universe Inside Out" (2003). It seems very strange indeed to be using the expression "more conventional hollow-earth theories"!

Cover of the July 5 1901 issue of *The Flaming Sword*

cosmology was described comprehensively by Koresh in his
book *The Cellular Cosmogony* (1898), but for our purposes is
perhaps best introduced – along with much else – by citing
in extenso a segment called "A Glance at Koreshanity" in the
July 5 1901 issue of the newspaper Koresh sponsored, *The
Flaming Sword*:

KORESHAN UNIVERSOLOGY is a complete system of the Science of the great Universe of life; and it involves the knowledge of the Creator and his creation. The name by which it is designated, in contradistinction to perverted Christianity, is KORESHANITY; and the new Religion must supplant Christianity, as Christianity supplanted Judaism. Koreshanity has come to fulfil the hope of the world in the liberation of humanity from the curse, in the establishment of the Kingdom of God in earth, the introduction of the New Era of Light and Life, of universal harmony and happiness.

What does Koreshanity teach? We present a brief summary of the System – a few cardinal points, which will serve to suggest the great scope of the System in its completeness. It is the antithesis of all modern theories, of all schools of thought. It is the climax of all mental progress, the ultimate and absolute truth of Being and Existence; it is the revelation of all mystery, the uncovering of the occult; the true explanation of all phenomena, the scientific interpretation of Nature and the Bible.

COSMOGONY. – The universe is a cell, a hollow globe, the physical body of which is the earth; the sun is at the center. We live on the inside of the cell; and the sun, moon, planets, and stars are all within the globe. The universe is eternal, a great battery, and perpetually renews itself through inherent functions, by virtue of which it involves and evolves itself.

ALCHEMY. – The Science of Alchemy is the Philosopher's Stone, the Key to the mystery of life. Chemistry is false; Alchemy is true! Matter and energy are interconvertible and interdependent; they are correlates; matter is destructible; the result of its transmutation is energy. Alchemy is the key to the analysis of the universe.

THEOLOGY. – God is personal and biune, with a trinity of specific attributes. God in his perfection and power is the God-man or the man-God, the Seed of universal perpetuity. Jesus the Christ was God Almighty; the Holy Spirit was the product of his transmutation, or the burning of his body.

MESSIANIC LAW – The coming of the Messiah is as inevitable as the reproduction of the seed. The divine Seed was sown nineteen hundred years ago; the first fruit is another Messianic personality. The Messiah is now in the world, declaring the scientific Gospel.

REINCARNATION is the central law of life – the law of the resurrection; reincarnation and resurrection are identical. Resurrection is reached through a succession of re-embodiments. One generation passes into another; the millions of humanity march down the stream of time together.

THE SPIRITUAL WORLD. – Heaven and hell are in humanity, and constitute the spiritual world; the spiritual domain is mental, and is in the natural humanity, – not in the sky.

HUMAN DESTINY. – Origin and destiny are one and the same. The origin of man is God, and God is man's destiny. God is the highest product of the universe, the apex of humanity. Absorption into Nirvana is entrance into eternal life – in the interior spheres of humanity, not in the sky or atmosphere.

IMMORTALITY IN THE FLESH. – Koreshanity declares and defines the laws of immortality, and its attainment in the natural world. The first step is recognition of the Messiah and the application of his truth. KORESH was the first in modern times to announce the possibility of overcoming death in the natural world, in the flesh.

CELIBACY. – The saving of human life consists in the conservation and appropriation of life in humanity. To become immortal, one must cease to propagate life on the plane of mortality. The standard of Koreshan purity is the virgin life of Jesus the Messiah. The Central Order of the Koreshan Unity is Celibate and Communistic. Celibacy obtains in the central nucleus, never in the world at large.

PSYCHOLOGY. – Koreshanity points to the basis of all psychic phenomena – the human brain. It explains the phenomena of spiritism, mental healing, etc., and teaches the science of the relation of mind and matter.

THE BIBLE. – The Bible is the best written expression of the divine Mind; it is written in the language of universal symbolism, and must be scientifically interpreted. Koreshanity demonstrates the truth and scientific accuracy of the Scriptures, and proves its astronomy, alchemy, theology, ethnology, etc. There is no conflict between the Bible and genuine Science; the Bible and the natural universe must agree in their expression of the divine Mind.

COMMUNISM. – Koreshanity advocates communism, not only of the goods of life, but of life itself. It has not only the scientific theory of communism, but is practically communistic in the relations and affairs of its own people. In this it corresponds to the primitive Christian church, where all things were held in common. The bond of the true communism is the true religion, and the central personality of the divine communism is the Messiah.

KORESHAN SOCIALISM. – Our Social System is patterned after the form of the natural cosmos; that form is the natural

expression of the laws of order. We demonstrate the fallacy of
competism; advocate the destruction of the money-power; the
control of the products of industry by the government, and the
equitable distribution of the goods of life. Koreshanity will
abolish wage slavery, and make it impossible for men to accu-
mulate wealth and impoverish the people.

    CHURCH AND STATE. – The true form of government is the
divine Imperialism: the unity of church and state; such will be
the Kingdom of God in earth. The Koreshan Government is
the unity of the empire and the republic, involving the principles of all present forms of government, which are but fragments of the perfect system which existed in ancient times – in
the Golden Age of the past. The government of the universe
is imperialistic; and humanity will constitute a unit only when
every class is emplaced at rest and liberty as are the strata,
stars, and spheres of the physical cosmos.

It seems that Koresh's ideas on the nature of the universe
came about when, while still living in New York State, he
became fascinated by the force of electricity, setting up for
himself an "electro-chemical laboratory" in which to
conduct experiments that might advance his knowledge
beyond that of stodgier researchers. According to his own
account, it was in 1869 that these experiments – with the
assistance of the Godhead's female aspect, or Divine
Motherhood – led him to the breakthrough realization from
which sprang the rest of the Cellular Cosmogony:*

> Form is a fundamental property of existence; therefore, that
> which has no form has no existence. Limitation is a property
> of form. The universe has existence; therefore it has form,
> hence it has limitation.

Working from the outside of Koresh's universe in towards its
centre, we have first a shell of gold; concentrically within
this follow six shells of other metals, five mineral shells, and
then five shells of rocky strata, on the innermost of which we
dwell. All around us is an atmosphere of air. Above that –
i.e., inward towards the middle – is a second atmosphere,

---

*Some cruel commentators have suggested that, rather, Koresh
received an electric shock which damaged his brain.

this time of hydrogen; and within *that* there's a third atmosphere, surrounding the sun at the universe's centre and made of a stuff called aboron.

That there are these three atmospheres is of prime importance in explaining our astronomical observations and more generally our experience that the universe is very much other than Koresh described. If we could directly see the sun – the heart of the electromagnetic battery that is the universe – we would observe that it takes the form of a helix that rotates once every 24 hours.* But of course we *don't* see the sun directly, because the light rays reaching us from it travel in curved paths, being refracted at each inter-atmosphere boundary; as a result of these refractions ("focalizations"), in different parts of the world the sun appears in different places in the sky – higher or lower from the horizon, etc. What we really see when we think we're looking at the solar disc is the relevant place where the light rays are being "focalized" at the interface between hydrogen and air on their way to reaching our eyes.

Only very great brightnesses can be seen through the combined three atmospheres – otherwise we'd be able to see the lights of the cities on the far side of the world, which of course we cannot do.†

You may require a stiff drink around now, because there is more.

One side of the sun is bright, and that gives us daytime. The other side of it is dark, but there are myriad tiny light sources on it – little leaks of the sun's radiant energy, if you like. We don't see these as a cluster of lights because, once again, of the refraction effects within the three atmospheres; instead we see the points of brilliance as if they

---

* It also has a precessional movement; hence the apparent precession of the equinoxes.

† The Nazis did, however, toy with the notion that they might be able to fire rockets through the atmospheres to strike at targets in the antipodes. In 1933 the Magdeburg City Council asked a team of rocket scientists including the young Wernher von Braun (1912–1977) to explore this possibility. Funds ran out before the scientists had been able to build a rocket capable of travelling more than a few hundred metres.

Koreshan cosmology "explained" diagrammatically in Teed's book
The Cellular Cosmogony (1898)

were scattered all over the sky, and we call them stars.
Sometimes the refractions are a little diffused (small-scale
atmospheric turbulence perhaps?), and we see what we
call nebulae. Those other hazy patches of light in the
night sky, comets, are a little harder to explain because they
move: here Koresh had to postulate the existence of a belt
of glittery crystals surrounding the sun.

The planets and the moon are even harder to explain, and further Koreshanite physics must be invoked. The sun emits gravic rays. These are not refracted at the atmospheric boundaries but travel straight through; they also travel straight through anything on the inside surface of the hollow globe – ourselves and our environment included – and keep on emanating outward through the rocky, mineral and metallic strata unless they hit some impermeable (impermeable to gravic rays) obstacle. Such obstacles are a number of discs floating among the metallic shells of the outer universe. On meeting one of these, the gravic rays are transformed and reflect back upwards through the strata and the atmospheres towards the sun. In the hydrogen atmosphere, when the inward-travelling levic rays collide with outward-travelling gravic rays there's a reaction that emits light. (Think of a matter–antimatter reaction, but controlled and very, very considerably toned down.) The light rays from these sources bat around among the atmospheres and refractive boundaries like a wasp in a jamjar before being seen by us as the tiny, slowly moving illuminated discs that we call planets.

Matters are yet more complicated in the case of the moon, whose image is formed in our skies by a sort of levic-ray reflection of all the outer shells – rocky, mineral and metallic. Hence the fact that the lunar disc looks all pock-marked and, well, crusty.

The gravic rays have another function. Disciples of Isaac Newton believe we're held down to the ground by the force of gravity. Those who've had their eyes unveiled by Koresh realize there's no such thing as gravity – it's merely an illusion – and that what is pinning us down is the rain of gravic rays from above.

In light of all this, it might seem astonishing Teed could be interested in anything so pedestrian as experimental testing of his theory. However, when the opportunity came along he seized it. One of the recruits to the Koreshan Unity was writer and journalist Ulysses Grant Morrow (1864–1950). In 1886 he proposed an experiment much like those done in England earlier on the Old Bedford Level (see pages 33ff). On the Old Illinois Drainage Canal, convenient to where the Koreshan Unity was then based in

Chicago, Morrow placed a target image 45cm above the water level and then looked at it from a point 8km distant along the canal through a telescope a mere 30cm above the surface. According to orthodox models, the curvature of the earth should have ensured that the target image was invisible through Morrow's telescope – indeed, that it was some 2.75m below the horizon he could see – but Morrow insisted he could observe it quite clearly. This didn't prove the water's surface – and thus the earth's – was concave, as required by Cellular Cosmogony, but it certainly seemed to prove it wasn't convex.*

Other experiments followed. In one, done on Lake Michigan, it was observed that the hulls of distant ships were indeed obscured if you looked at them through low-power binoculars. Before globists might think to crow, however, further tests were done using higher-power telescopes, and now the distant hulls could be clearly seen! Alas, it was pointed out in due course that this phenomenon could be explained by the conventional refraction of light, as could the observations on the Old Illinois Drainage Canal, and Morrow – while disputing this – sought other means of testing.

After the Koreshan Unity had made the move from Chicago to Florida in 1894, Morrow had access to long straight beaches, and so the idea for a completely different type of experiment occurred to him. This depended on a device that he invented called the rectilineator.

Start with a thick, straight plank of mahogany 3.66m long. At each end fix a 1.22m plank at its precise centre and at a precise right angle, so that what you now have looks like a capital "I" with long serifs. Join the opposite corners together with taut steel bars to preserve those precise right angles; to the same purpose, add wooden cross-braces.

Make several of these units, and add fittings such that

---

* I confess to a failure to understand why, assuming a concave surface as decreed by Koresh's cosmology, Morrow didn't see the target image floating 2.75m *above the water surface*, and why he wasn't concerned that he didn't. Of course, Koreshanite cosmology made such play with the extraordinary refraction capabilities of light that perhaps *whatever* Morrow observed would have been taken as "proof".

they can be bolted together very meticulously to ensure the line of one longitudinal beam can be exactly continued by the next.

Now you need vertical posts correctly fitted with attachments to support your beams horizontally. Once you have done all this – and more besides that we don't have space to describe here – and assuming you have a long, flat, empty stretch of beach like the one* near Naples, Florida, you are ready to perform Morrow's experiment.

The intent was to use a string of rectilineators, each affixed to its neighbours with precision, in order to produce an exactly straight horizontal line over a distance of some kilometres; with this exact straightness, one could measure to see if the earth's surface curved beneath the line – and, if so, whether it curved upwards or downwards. The reason for doing this on a beach was that, through use of caissons (and a bewildering array of calculations to allow for tides and the like), a precise tally could be gained of any discrepancies between the horizontal as measured by the rigid rectilineators and the horizontal as measured by the water's (i.e., the earth's) surface.†

Morrow's results, rigorously summarized in the book *The Cellular Cosmogony*, showed quite clearly that the earth was curving upwards! In other words, we were living on a concave surface, which could only mean that the Koreshanite cosmology – or something very like it – was correct! Even more impressively, the measured curvature over the 6.5km of the experiment accorded neatly with the concave earth having a circumference of about 40,000km, just as theory predicted. It is no wonder that the two men, Morrow and Koresh, and their supporters sounded somewhat triumphalist in their writings about the outcome.

How could this have occurred? In his excellent analysis of the experiment (see Bibliography), Donald Simanek

* Not so empty these days!

† Rectilineators were pieces of precision engineering and cannot have been cheap or easy to make. The Koreshan Union could afford neither the cash nor the effort to make hundreds of them. Instead they made a relatively small number, perhaps a dozen or so, and, once all had been erected, removed the first to join it onto the advancing end of the line.

Rectilineators in use in Morrow's experiment

shows convincingly that a systematic error arose in
Morrow's experiment because of assumptions made about
the rigidity of the rectilineators. Beams of mahogany are
heavy, and their ends sag when the beams are supported
horizontally – especially if there are further weighty beams
attached to the ends. Morrow did make some effort to counteract this by supporting each 3.66m beam in two places,
but he seems to have underestimated the degree of the
problem because he didn't bother to calculate the position
of those two supports precisely. The wooden cross-braces
intended to assist in retaining perfect right angles
contributed to the effect, being themselves not without
weight. And Morrow seems to have overestimated the efficacy of the steel cross struts in maintaining rigidity.

Mahogany beams don't sag by very much; with the
instruments available to him, Morrow couldn't possibly have
detected the sag over the length of a single rectilineator, or
even over the length of all his rectilineators joined together.
But his experiment involved the use of about 1780 (virtual)
rectilineators, and if you multiply by that number the very
small amount of droop in a single rectilineator you arrive at
something very detectable indeed.

As to the matching of the "curvature" with the circumference of the earth? Simanek proposes, almost certainly

correctly, that this was a result of unconscious experimental bias. Morrow began to observe what he thought was proof of the concavity of the earth's surface, and his enthusiasm to prove Cellular Cosmogony correct took over, despite his own conviction that he hadn't allowed it to, when it came to the mathematical details.*

The idea of our living inside a hollow earth did not quite die out with Teed and his disciples. In 1947, according to a *Time* magazine account in July of that year, the Argentinian amateur astronomer Antonio Duran Navarro derived it, or something very like it, seemingly independently. The universe, Navarro announced, is some 13,000km in diameter and contained entirely within the earth. In 1983 the mathematician Mostafa Abdelkader published a paper containing a mathematical model of the local part of the universe that would work just as well for hollow-earth and orthodox explanations; he preferred the former. And so on. A number of websites exist that promote Koreshanite or similar cosmologies; some are quite probably sincere, and some look impressively scientific, with equations and Greek letters and stuff. Oh my.

———————⟨✿⟩———————

Some people really believe the velocity of light isn't what we think

it is. And there are others only too willing to cash in on such reasonless misbelief, as the Nazis were when, eager to discredit "Jew science" – like Einstein's Relativity – they promoted the crackpot cosmology of an Austrian mining engineer called Hanns Hörbiger (1860–1931). In his theory, known variously as the World Ice Theory, the Cosmic Ice Theory and Glacial Cosmogony (*Glazial-Kosmogonie*), the only star in the universe (using our conventional meaning of the word "star") is the sun. All the rest are

---

* Unconscious experimenter bias is a well documented problem among even the most seemingly objective scientists. A number of examples are discussed in my book *Corrupted Science* (2007).

just chunks of ice, gleaming by reflected light. Similarly, the earth is the only planet not to be ice-covered; it shares the same fate as the other planets, though, in that it will be eventually devoured by the sun as all the smaller bodies of the solar system spiral in towards this fiery finale, being replaced by new planets on a kind of conveyor-belt principle.

At first glance the cosmology of Immanuel Velikovsky (1895–1979) might seem a trifle less divorced from the mainstream – and certainly it was less ambitious, being concerned merely with the solar system rather than with the entire universe. Velikovsky's real concern was a rewriting of ancient history/prehistory, as reflected in the titles of his books: *Ages in Chaos* (1952; published second but written earlier), *Worlds in Collision* (1950), *Earth in Upheaval* (1955), *Oedipus and Akhnaton* (1960), *Peoples of the Sea* (1977) and *Rameses II and his Times* (1978).* This revised history was marked by various catastrophes of global scale, many of them connected to the emission by the planet Jupiter of a comet that would eventually settle down to become the planet Venus, but not before swooping repeatedly around our own world and causing upheavals of the land and sea, rains of chemicals, etc. For a good quick survey of Velikovsky's cosmological ideas, and of all the reasons why they make no scientific sense, consult *Scientists Confront Velikovsky* (1977), edited by the distinguished astronomer Donald Goldsmith.

Don't get the impression that wilfully perverse cosmologies are exclusively a 20th- and 21st-century phenomenon. There were plenty of them in earlier periods too. For example, Lord Monboddo (see page 263) had the theory that all things, including celestial bodies, possessed minds. To be sure, inanimate objects like rocks and even planets had less sophisticated consciousnesses than, say, vegetables, which in turn had cruder minds than – ascending the ladder – animals, humans, angels and God, but they were neverthe-

* There were also two posthumous works, cobbled together from unpublished writings: *Mankind in Amnesia* (1982), containing the great man's thoughts on psychoanalysis, and *Stargazers and Gravediggers* (1983), recounting the hostility of science towards his theories.

less to some extent sentient. This panpsychic cosmology of
Monboddo's was much simpler – and therefore, he
believed, much more accurate – than that of Isaac Newton.
Whereas Newton had to work out the laws of gravity in
order to explain the elliptical paths of the planets,
Monboddo saw right through to the heart of the matter: the
planets orbited the sun basically because *they wanted to*. That
was the nature of the consciousness God had given them,
and to seek for any further explanation was futile.

Many of the more recent unorthodox cosmologists
seem to have been inspired by a quasi-instinctive repug-
nance towards Relativity. (Certainly this was a part of the
Nazis' espousal of Hörbiger's ideas, even though their
motives were primarily political/racist.) A fierce opponent of
Einstein's Relativity from the outset was the inventor Nikola
Tesla (see page 203), who maintained that Newton's physics
was far better than this newfangled nonsense – although
even Newton's physics was not nearly so good as the new
physics that he, Tesla, had worked out. Unfortunately he
never released any details of this marvellous physics, so to
this day no one knows what he was talking about. He did,
however, give a hint relating to it when, late in life, he
announced he'd discovered how to tap the hitherto unde-
tected energy source that powered the universe: ". . . a
source to which no previous scientist has turned, to the best
of my knowledge. The conception, the idea when it first
came upon me, was a tremendous shock." Sadly, though,
here again he omitted to reveal any of the particulars.*

Another early critic of Relativity was Arthur Lynch
(1861–1934), author of *The Case Against Einstein* (1932).
David Langford and Chris Morgan summarize Lynch's
arguments in their *Facts and Fallacies* (1981) thus:

> . . . the shifts of spectral lines produced by the Sun's gravity
> mean nothing because the Sun is a long way away. The bend-

---

* Tesla's lifelong habit was to invent by means of what we'd today call
thought experiments. Rather than build prototypes, test them, refine
the model, etc., he'd do all the building, testing and refining in his
mind. Once he'd finished an invention in this manner, he'd very often
just leave things at that. We thus have no guarantee many of the marvellous inventions he claimed would have worked if built.

ing of light as it passes the Sun means nothing because not all measurements of the amount of bending agree exactly. The shifting of the orbit of Mercury means nothing because Einstein fiddled his original equations so as to account for it.

Quite a few rungs up the ladder of respectability is the critique of Relativity offered by Herbert Dingle (1890–1978) in his book *Science at the Crossroads* (1972) and elsewhere. Dingle's difficulties with Relativity began with this question:

> If A is moving at great speed relative to B, B will notice that A's clock is running slow (according to Einstein). But, *so far as A is concerned*, it is B who is moving at great speed, and so A will notice that B's clock is running slow. Can two clocks *really* each be running slower than the other?

The short answer is, basically, yes. To any outside observer, C, it's perfectly obvious that *both* clocks are wrong, and so on. Of course, the short answer is a gross simplification. It took Dingle a long time to find someone who'd give him the long answer, and by then it seems he was so much in the habit of complaining Relativity didn't make sense that it was difficult for him to stop.

Needless to say, Charles Fort (see page 94) had his own unique quibbles with Relativistic notions. A cornerstone of Special Relativity is that nothing can be accelerated up to and beyond the velocity of light. Many unorthodox cosmologists have queried this statement. Fort's was by far the most fundamental question: has it yet been proved light has a velocity at all?

Dodging this issue, let's turn to other, less thought-provoking statements. Dino Kraspedon, who had it from the captain of a UFO, tells us the velocity of light in free space is infinite. W. Raymond Drake joins this chorus: a supporter of the ancient-astronaut hypothesis, he realizes the lightspeed restriction bodes ill for would-be interstellar visitors – but, he says, "this objection is overruled by the possibility that in the near vacuum of space light travels many times faster than in our own atmosphere and may approach infinity". No air-resistance to worry about out there, eh?

And then there's Jacques Bergier (1912–1978), a doyen

of unorthodox thinkers: "The speed of light represents only a convenient, not an absolute, speed limit, like the sound barrier, for instance." Of course, the similarity between the sound barrier and the light barrier goes no further than the word "barrier". Bergier continues: "In a material medium, in which light is braked, no object can be faster than light without breaking the Einsteinian laws. But in a vacuum . . ." Yes, yes, a spaceship can travel through a vacuum faster than light can travel through a brick wall, but can it travel faster than light can travel *through a vacuum*? – which velocity is, after all, the "velocity of light" referred to by Einstein. Or is Bergier thinking of our old friend air-resistance?

Harold Camping of Family Radio, concerned to argue that the earth is just a few thousand years old and aware of the problem that, if so, the light from distant stars wouldn't have had time to reach us yet, proposes the universe is just a few light years across; those very faint stars we see aren't distant, just small and dim. What Camping fails to recognize is that stars have to have a certain mass to be stars at all: below that mass their nuclear fusion reactions won't start.

An alternative explanation as to how light could travel massive distances in just a few years was offered by Barry Setterfield in *The Velocity of Light and the Age of the Universe* (1983). He looked at past calculations of the velocity of light, going back as far as those of Ole Rømer (1644–1710) and Jean Picard (1620–1682) in 1676, who proposed a figure of 299,270kps, +/–5%. This is a remarkably good piece of measurement, bearing in mind the equipment available to 17th-century scientists: the velocity of light is in fact 299,792.5kps. However, Setterfield seized upon that cautious "+/–5%" and elected to look only at the topmost value in the range: 301,300kps.

Let's get this straight: Rømer and Picard estimated the velocity as 299,270kps, which is *slower* than the now-known figure. Setterfield, however, chose to look instead at 301,300kps, which is *faster* than the true figure. Further, the known figure for the velocity of light falls smack bang within Rømer/Picard's range.

Whatever, using the 1676 and other figures Setterfield concluded that the velocity of light has been decreasing exponentially with time since the creation of the universe. Plugging in a date of 4004BC on his exponential curve, he was able to deduce that light would then have had a velocity in free space some 500 billion times greater than it currently is. Thus distant stars are *in fact* at most only a few thousand light years away, so long as you realize a light year in the past was a very much longer distance than a light year today.

There are further refinements. The velocity of light did not start decreasing until some little while *after* the creation of the universe – until the Fall, in fact. Before then, God obviously would not have permitted His universe to contain anything so imperfect as a decay of the velocity of light. Another tweak came from Setterfield when it was pointed out that light-velocity measurements from the 1960s gave values identical with those of the 1980s; these measurements were certainly accurate enough to have revealed any reduction. The explanation is that, for reasons unknown, the velocity of light stabilized in 1960 and has remained static ever since.

Another problem is that, if you play around with the velocity of light, all sorts of other physical properties of the universe change as well. Take Einstein's famous equation $E = mc^2$ and then consider what happens to the heat output of the sun if you increase the value of c. Forget about a factor of 500 billion: a mere doubling would be more than enough to make our world prettily incandescent.

Some people really believe they understand quantum physics. If Relativity is unpopular among the bogus theorists, the same cannot be said of the quantum theory. These days it seems the belief is rife that merely adding the word "quantum" to something makes it scientifically respectable or, if not that, anyway *sexier*. A visit to your local bookstore should confirm this. Titles abound like *Quantum Shift in the Global Brain: How the New Scientific Reality Can Change Us and Our World*

(2008) by Ervin Laszlo, described by its publisher thus:

> Ervin Laszlo presents a "reality map" to guide us through
> today's quantum world shifts. We need this map to understand
> what we must do during this time of great transition as old
> ways of thinking yield to new multi-dimensional realities.

All of which means exactly . . . what? Well, to paraphrase
Richard Feynman (1918–1988), if you think you understand
quantum physics, that proves you don't.

Even religious belief can be given a patina of "scien-
tificness" by addition of the magic word. In his *Quantum
Theology: Spiritual Implications of the New Physics* (1997)
Diarmuid O'Murchu interestingly distances theology from
religion, which he regards as being an artificial expression
of spirituality that has by now outlasted its usefulness – an
unorthodox proposition for a Christian priest! Instead, he
maintains that the proper study of theology is spirituality
itself, not its offshoot; and that in this sense *Homo sapiens*
has been "doing" theology since long before the
Agricultural Revolution. In support of this contention
O'Murchu is not afraid to make some sweeping generaliza-
tions about cultural norms prevalent tens of thousands of
years ago.

> Over the millennia – some seventy thousand years – we
> humans lived in a spiritual ambience. We sought and discov-
> ered meaning in the events of daily life. We sensed the fright-
> ening, yet benevolent, power of the divine in the rhythms of
> nature, in the changing seasons, in the warmth of sunshine, in
> the light of the moon, the destruction of storm and thunder.
> The entire universe was alive with potential meaning,
> perceived for over thirty thousand years as a Divine Mother of
> prodigious fertility and nurturance . . .

. . . which is all very possible, but for which we have not one
scintilla of proof.

Critics, this one included, have complained O'Murchu's
text presents a thesis that might best be described as diffuse,
marked by an apparent reluctance to define what Quantum
Theology actually *is*. It seems he's much taken with the
quantum notion that there can be no such thing as an inde-

pendent observer of an event: the experimenter is an active part of the experiment. Taking this further: as a part of the evolving universe, we play an active role in that evolution; we are the universe's ongoing co-creators alongside the deep spiritual force which some call God.

One immediately noticeable problem with O'Murchu's theorizing (aside from the fact that he or his copy editor can't spell the word "holistic", giving it throughout as "wholistic") is that he seems to lack the mental filter which might indicate to him that there's a profound qualitative difference between the quantum speculations of genuine physicists (Paul Davies, John Barrow, John Polkinghorne, etc.) and those of various New Age pseudo-physicists (Deepak Chopra, Danah Zohar, etc.) who have latched onto "quantum" as the next great exploitable "mystery". The result is that, while often O'Murchu's expositions of aspects of quantum theory are quite good, not infrequently they descend into a decided quirkiness.

In an appendix O'Murchu presents a dozen "Principles" which together might be taken to compose Quantum Theology; each "Principle" is qualified by a few "New Elements". As an indication of how difficult it can be to get one's head around these "Principles", here's the first:

> Life is sustained by a creative energy, fundamentally benign in nature, with a tendency to manifest and express itself in movement, rhythm, and pattern. Creation is sustained by a superhuman, pulsating restlessness, a type of resonance vibrating throughout time and eternity.

Swathes of New Age and related books try to capitalize upon a perceived similarity between the Hindu vedas and quantum theory/particle physics. The rot (in more than one sense of the word) started with Gary Zukav's book *The Dancing Wu-Li Masters: An Overview of the New Physics* (1979), a work that received colossal praise from nonscientists when it was published but today tends to seem quaint. A somewhat earlier and weightier book along similar lines was *The Tao of Physics: An Exploration of the Parallels between Modern Physics and Eastern Mysticism* (1975) by Fritjof Capra (b1939). As Richard Dawkins (b1941) cruelly characterized all such essays, "Quantum physics is deeply mysterious and incomprehensible. Eastern spirituality is deeply mysterious and incomprehensible. Therefore they must be saying the same thing."

Modern mystics are fond of talking about higher – astral – dimensions, while modern physicists and cosmologists seem increasingly to be drawing upon the notion of there being further dimensions than the ones to which we're accustomed. Needless to say, there's not a huge amount of similarity between the two sets of concepts both described by the single word "dimensions"; likewise needless to say, numerous New Agers have jumped upon the coincidence of terminology and made much of it.

This is not a new practice. Recently used as the basis for the animated movie *Flatland* (2007), the novella *Flatland: A Romance of Many Dimensions* (1884) by Edwin A. Abbott (1838–1926) is a piece of fantasy/science fiction that explores our perceptions of the dimensions by positing the existence of a two-dimensional world, the Flatland of the title. Denizens of Flatland can accept the existence of a zero-dimensional world (Pointland) and a one-dimensional world (Lineland), but are mystified by the periodic arrival in their own world of a three-dimensional entity, a visitor from Spaceland. Our hero, a middle-class Flatlander, goes to the other worlds, including Spaceland, and posits to his friend there that there must be other, higher dimensions than the Spacelanders can perceive. They reject this notion derisively.

Abbott intended his work to be, as well as entertainment, a demonstration of mathematical concepts. Some of his readers were entranced by the idea of further dimensions and speculated further. Among these was the British maths teacher C.H. Hinton (1853–1907), who published his own novel, *An Episode on Flatland, or How a Plane Folk Discovered the Third Dimension* (1907).* Hinton was fascinated by the fourth dimension, for mystic rather than mathematical reasons, and posited four-dimensional solids (tesseracts) called hyperspheres and pentahedroids, all three of which names he coined. (He also named the oppos-

---

*He also fled to Japan and later the US after being convicted of bigamy.

ing directions in the fourth dimension – the equivalent of left and right, or up and down – as "kata" and "ana".) The fourth dimension was in those days visualized as being a further spatial dimension rather than, as Einstein's Relativity would establish, the dimension of time.

In an echo of what had happened after publication of Abbott's book, Hinton's ideas, which were imaginative but not crazy, were embraced by some readers of an occultistic bent. Among these was the Russian-born US occultist P.D. Ouspensky (1878–1947), author of *The Fourth Dimension* (1909), *Tertium Organum: The Third Canon of Thought, a Key to the Enigmas of the World* (1912) – modestly intended as the conclusion to the trilogy started by Aristotle with *Organon* and continued by Francis Bacon (1561–1626) with *Novum Organum* (1620) – and of *A New Model of the Universe: Principles of the Psychological Method in its Application to Problems of Science, Religion and Art* (1930), all of which are as grandiose in ambition as the subtitles of the latter two might suggest. Motion in the fourth dimension, according to Ouspensky's earlier works, affects atoms and molecules in ways that can be perceived in three-dimensional contexts such as snowflakes and growth. Later he accepted the Einsteinian notion that the fourth dimension was time. Ouspensky decided time was spiral in nature (this seems to relate to his oddball version of reincarnation, whereby at death we go back to the moment of our birth and live the same life all over again), and therefore must have two dimensions of its own – to bring up a grand total of six. I think.

Some people really believe time is in disarray.

If the fourth dimension is time, is it possible – ask many unorthodox theorists – that scientific ideas about it are as haywire as all other scientific ideas?

The UK aeronautics engineer* John William Dunne

---

* Among Dunne's many achievements in this sphere was that, in 1906–7, he designed and built the UK's first military aeroplane.

(1875–1949) probably did more than any other person – *pace* the sustained efforts of countless psychic researchers – to imprint on public consciousness the ideas that (a) many dreams seem to display elements of precognition and (b) the study of this phenomenon is a perfectly respectable scientific enterprise. In his bestselling book *An Experiment with Time* (1927) Dunne presented readers with a fair number of his own purportedly clairvoyant and precognitive dreams, collected over a period of decades, dreams which suggested to him that our whole concept of the way time "works" is fundamentally flawed. In response to this conclusion he generated a new theory of time, which he called Serial Time. In successors such as *The Serial Universe* (1934) he expanded this notion.

In detail his hypothesis is as close to incomprehensible as makes no difference. In the most approximate terms, he suggested there exists a Time 2, and that it's against Time 2 that we measure the flow of Time 1, the dimension we're normally referring to when we use the word "time". If we could somehow manoeuvre ourselves into a Time 2 existence, we could see the whole of Time 1 stretching ahead of and behind us as if a static plain. In other words, although we think of the past and the future as inaccessible to us, in reality the whole of time – past, present and future – exists unchanging all around us, if only we had the perception to see it that way: our consciousness is what imposes upon us the experience of time as a linear progression. In other words, our existence in Time 1 is analogous to our walking along a road and wondering what's around the next bend; the fact that we can't see what's round that corner obviously doesn't mean the scene doesn't exist – in fact, if we were able to fly we could see it easily. Of course, as creatures of Time 1, we don't have the ability to see what's around the bend of the future . . . except sometimes, in special circumstances, such as when we dream. In general, being in the dreaming state relaxes some of the inhibitions consciousness places upon our perceptions – hence the (supposed) experience of precognitive dreams.

The dreams Dunne had which struck him as precognitive are of varying levels of interest. In one instance, in 1901 while he was on the Riviera, he dreamed he was in the

Sudan, possibly at Fashoda – where, in 1898, just a few years before Dunne's dream, the UK and France had almost come to war over an oasis neither of them actually wanted. The dreaming Dunne saw three men approaching from the south; they were dressed like the soldiers with whom he had recently been trekking in South Africa during his service in the Boer War. He asked them what they were doing so far north. They replied that they'd come all the way from the Cape, and that one of them had almost died of yellow fever *en route*.

The next day, Dunne tells us, he learned from his newspaper about the Cape-to-Cairo Expedition, which he'd had no idea was in progress. According to the newspaper report – which was of course out of date by the time Dunne read it – the expedition had reached Khartoum, not far from Fashoda. Later he found a report that three of the white expeditioners had died during the journey, although they had succumbed to enteric rather than yellow fever.

This dream's status as precognitive does not seem nearly as strong as Dunne obviously thought it was. Although he was steadfast in his belief that he knew nothing beforehand of the Cape-to-Cairo Expedition, it's obviously possible he'd overheard something about it and forgotten. His dreamed information about a death on the journey was wildly different from the actuality. Khartoum and Fashoda are close to each other, but that shouldn't obscure the fact that the dream quite plainly specified the wrong place.

The most famous of his supposedly precognitive dreams was that involving the Mont Pelée disaster of 1902. Now back in South Africa, he dreamed one night he was on a volcanic island about to erupt. With the ghastly lesson of Krakatoa very much on his mind, he began to make efforts to have the 4000 unsuspecting inhabitants of the island shipped off in time. His dream then became an anxiety dream as he tried to get French official after French official on a neighbouring island to send ships to enable the evacuation. He awoke in a sweat while clinging to the horses of the carriage of one such official, who – deaf to Dunne's cries of "Listen! Four Thousand people will be killed unless . . ." – was irritatedly pointing out that it was dinnertime and telling Dunne to call back tomorrow.

Some time passed before Dunne next saw a newspaper. When he finally did so, the first headline to meet his eyes was:

VOLCANO DISASTER IN MARTINIQUE
Probable Loss of Over 40,000 Lives

The report beneath was of the Krakatoa-style eruption of Mont Pelée. But there were some peculiarities in terms of the complete experience having been precognitive. First, although it has been widely stated that the newspaper got the casualties figure wrong, in fact the figure seems to have been only mildly inflated, if at all.* Whatever the accuracy, Dunne misread the figure in the headline: he saw it as "4,000" rather than as "40,000", and it was only some 15 years later that he realized his error. What prompted the misreading was that his mind was on the figure that had been so important in his dream.

Dunne's own explanation for both this dream and the Fashoda one was that what his unconscious mind had predicted were not the real-world events but *his experiences of reading about the events in his newspaper*. Of course, those discoveries were of little significance in the overall scheme of things, but they had a considerable effect *on him*. The only trouble with this line of reasoning is that it's circular: had he not had the dreams, then reading about the events wouldn't have had (one assumes) much of an impact on him at all.

It's because of apparent paradoxes like this that scientists generally decline to believe there can be such a phenomenon as precognition. People who're convinced of the existence of precognition argue that, if science can't easily incorporate the idea of information passing from the future to the past, then it's science itself that's at fault, not the phenomenon.

---

*The island's main city, St Pierre, was almost completely wiped out, with only a very few survivors. As the city's population had been about 30,000, with at least a few thousands more having fled there from outlying areas during the preceding fortnight or so as the volcano grumbled, and as casualties were not restricted to St Pierre, a total death toll of 40,000 seems not implausible.

Dunne went on to argue, in books like *The New Immortality* (1938) and *Nothing Dies* (1940), that, if the two forms of time – the linear and the omnipresent – exist simultaneously (as it were), then we ourselves must likewise exist on both levels. But if we have an existence in Time 2, that existence must be eternal; in other words, at a level we normally cannot perceive, we're immortal.

These notions of Dunne's are not without precedent in various mystical traditions. Perhaps the most striking parallel is with the Australian Aborigines' idea of the Dreamtime, which, like Time 2, exists "simultaneously" through past, present and future. The Dreamtime is the "real" time, according to this tradition; linearly experienced time (Time 1) is merely a product of human perception, and thus essentially an illusion. Quite how the behaviour of animals – which are palpably capable of telling past from present – fits into the Dreamtime concept or Dunne's hypothesis is something I've been unable to establish.

Perhaps Dunne's most famous literary champion was the novelist J.B. Priestley (1894–1984), who incorporated some of Dunne's notions into his play *Time and the Conways* (1937). In his lengthy essay *Man and Time* (1964), Priestley humorously recalls: "[A]t his own suggestion, he explained his ideas to the cast of my play *Time and the Conways*, a cast that always played well but never better than when they were pretending to understand what he was telling them . . ." Indeed, Priestley devoted a whole chapter of that book to discussing Dunne's theories, declaring that

> About his originality, his entirely new approach, the audacity and sweep of his conclusions, there can be no question. He opened a way, and, whatever my reservations may be, I think it is the right way. It is right, to my mind, because he rejects the idea, almost a dogma now, that our lives are completely contained by chronological uni-dimensional time, without becoming other-worldly.

In this passage the old maestro seems to be letting his wishes leap ahead of his intellect (if we accept for the moment that Dunne might have been right to reject the conventional picture of time being "chronological uni-dimensional", this is by no means good enough reason to swallow the rest of his theories unquestioned), but in fact Priestley isn't being so foolish. Later he explains why he can't accept the fundamentals of the Serial Time hypothesis:

> As soon as Dunne reaches Part IV of *An Experiment with Time*, leaving his dream experiences and turning to theoretical exposition, he seems to me to go wrong almost at once. He goes wrong because he insists on using the concept of Time in two different and contradictory ways. . . . What he does is to spatialize Time and to treat it as movement, both at once. . . . I have no objection to a spatial view of Time. Such a view is inevitable, sooner or later, if only as a convenience in argument. What we cannot legitimately do is to dodge between Time lying flat and motionless as an extra dimension of Space, and Time in its older role as movement. We must make one approach or the other; we cannot hop between the two. But this is what Dunne did from the first: it is the basis of his main argument.

Dunne's speculations about time seem positively pedestrian alongside those of the UK philosopher John McTaggart Ellis McTaggart (1866–1925). Based at Trinity College, Cambridge, for almost all his academic life, McTaggart demonstrated to his own satisfaction in his two-volume work *The Nature of Existence* (1921–7) that time, matter, space and God do not exist, but instead arise from our misperceptions. Those proofs are far too complex for us to go into in detail here, but we can take a quick look at one of his demonstrations of the nonexistence of time.

According to McTaggart, no event can exist simultaneously in past, present and future; these three characteristics of an event would be incompatible. However, it is incontrovertible that all events are *related to* past, present and future: events occur because of prior causes, and they have later consequences. But if the three characteristics are incompatible, they cannot be related to each other through the event. As we can directly perceive the event, it cannot be the false component of this apparent paradox. Thus the inescapable conclusion: the flow of time is an illusion born of our misperceptions.

A further consequence of this inference is that the universe cannot have been created by a god. The act of creating the universe would obviously imply a causal relation between the god and the universe; but causal relations require the cause to *precede* the effect – something that's impossible in a reality where time doesn't exist. Similarly, the notion that, even though the universe wasn't created by a god, it might be governed by one is untenable: such a god would have to cause events, and as we've just seen this would be feasible only if the flow of time existed. Perceptions by human beings that a god exists are mere value judgements; as we know from exhaustive experience in other areas, human value judgements are as likely to be wrong as right – if not more so!

How are we to reconcile these notions with what we perceive (mistakenly) as reality? In his 1908 paper "The Unreality of Time" (in the journal *Mind: A Quarterly Review of Psychology and Philosophy*) McTaggart explains:

> And so it would seem that the denial of the reality of time is not so very paradoxical after all. It was called paradoxical because it seemed to contradict our experience so violently – to compel us to treat so much as illusion which appears *prima facie* to give knowledge of reality. But we now see that our experience of time – centring as it does [on] the specious present – would be no less illusory if there were a real time in which the realities we experience existed. The specious present of our observations – varying as it does from you to me – cannot correspond to the present of the events observed. And consequently the past and future of our observations could not correspond to the past and future of the events observed. On either hypothesis – whether we take time as real or unreal – everything is observed in a specious present, but nothing, not even the observations themselves, can ever be in a specious present. And in that case I do not see that we treat experience as much more illusory when we say that nothing is ever in a present at all, than when we say that everything passes through some entirely different present.

Yet a further consequence is that, since we exist in a timeless universe – i.e., we are a part of that universe – then we too must be timeless: we reincarnate indefinitely. This aspect of McTaggart's philosophy has been enthusiastically embraced by the actress Shirley MacLaine (b1934) in such New Age works as *Out on a Limb* (1983).

Some people really believe in psychic

physics. Also known through different translations as od or odyle, the odic force was the brainchild of the German industrial chemist Baron Karl von Reichenbach (1788–1869). Named for the Norse god Odin, this force was a power or emanation somewhat similar to the animal magnetism posited by Franz Mesmer (1734–1815) or the orgone energy hypothesized much later by Wilhelm Reich (1897–1957). It permeated the entire universe, being emitted by all things, whether stars or planets, organic or inorganic entities, and most pertinently by the human body, in which latter case it might be thought of as a variant on the aura concept. Human emissions of odic force could be detected only by sensitives – not psychics but simply people with the ability to pick up odic emanations, which they saw as clear "flames" in pure colours streaming from fingertips or mouths or foreheads, from magnetic poles, from recent graves, etc. This ability seems to have been more of a nuisance than anything else, as von Reichenbach discovered during the experiments he ran from 1839 onwards with some two hundred such sensitives; with it came, for example, a sensitivity – usually unpleasant – to various metals, notably brass. Sightings of supposed ghosts in cemeteries are (in effect) optical illusions; they occur because decayed corpses emit "odic light", which sensitives can see. Another feature among the sensitives was a dichotomy between right and left sides: they reacted differently to people depending upon whether those people were to the individual sensitive's left or right, an attribute that significantly affected their sleeping arrangements with spouses.

Von Reichenbach didn't regard his researches as in any way pertaining to the psychic; he seems to have hoped his odic force would be recognized as an integral component of the universe in the same way as other forces like electromagnetism and gravity. Ironically, while his contemporaries tended to give his theories short shrift, it was precisely because of the putative psychic connection that his reputa-

tion suddenly revived some while after his death, when the
Society for Psychical Research became interested, forming a
Reichenbach Committee. Although the committee had
difficulty finding appropriate sensitives, they did find a few.
Elsewhere, Samuel Guppy (1795–1875), author of books
like *Mary Jane, or Spiritualism Chemically Explained* (1863)
and husband of not one but (successively) two mediums,
believed that manifestations at seances could be explained
in terms of odic vapours being emitted from the medium's
body.

Elsewhere in the wild universe of psychic physics, in
Posthumous Humanity, or A Study of Phantoms (1887), trans-
lated by the Theosophical stalwart Henry S. Olcott
(1832–1907), Adolphe d'Assier (b1828) attempted to bring
science to bear on much of the field of what's now generally
termed the occult or paranormal, including such exotica as
vampires and werewolves. Here he is on the physics whereby
wraiths eventually dissipate, so that they haunt no more:

> The molecules of the phantom's tissues disintegrating from
> each other, there comes a day when it has no further
> consciousness of itself. In slow agony, it becomes weak.
> Tumultuous at first, it grows less and less so, as the shade
> suffers from cosmic ailments, until definite annihilation
> occurs.
>
> But, while young and lusty, ghosts produce mechanical
> effects as great as if they were of large bulk, as their noisy
> habits testify. In lieu of stones, ghosts throw their duplicates
> which have the same effect and obey a rigid formula: its life
> force at the moment of fall is equal to half its bulk multiplied
> by the square of its velocity.

D'Assier's book is a fount of that curious kind of wisdom you
wonder why you never had before. Here's another example:

> The accumulation of specters of the different tribes of the
> terrestrial fauna, heaped at the surface of the globe since the
> first geological epochs, would render the air irrespirable. We
> could not move, in a dense atmosphere of ghosts.

————⟨⟩————

Some people really believe that Charles Hoy Fort (1874–1932) saw to the heart of matters and brought forth into the sunlight those strange discrepancies – the "damned" – that even today scientists would conspire to sweep under the carpet. In his sympathetic biography *Charles Fort: Prophet of the Unexplained* (1970) the science fiction writer Damon Knight (1922–2002) put it like this: "His data were like unwanted children deposited on scientists' doorsteps."

After bumming around the world for a few years in his late teens and early 20s, Charles Fort tried to make it as a novelist, although only one of the ten novels he wrote before turning to his curious variety of nonfiction, *The*

Charles Hoy Fort, who cast an unrelenting spotlight on mysterious events that science couldn't explain . . . or perhaps just on excesses and errors of yellow journalism

Outcast Manufacturers (1909), saw print; it received a friendly critical response but sold poorly. An inheritance he received in 1916 enabled him, nevertheless, to devote the rest of his life, which he spent in New York aside from a brief excursion (1924–6) in London, to writing. His four main books – *The Book of the Damned* (1919), *New Lands* (1923), *Lo!* (1931) and *Wild Talents* (1932) – are well known and widely in print. Two earlier books, which he called *X* and *Y*, remained unpublished despite the best efforts of the novelist Theodore Dreiser (1871–1945), acting as Fort's literary agent, to interest publishers in them. (A planned *Z* was never written.) It seemed Fort was ruthless with his manuscripts, and when it looked evident to him that these two books would fail to find a publisher he discarded his written texts of them – as he did those of various unpublished novels. (Of course, if he'd waited until after the publication of his other books, there's every chance a publisher would have been happy enough to buy *X* and *Y* as well.) All we have left of these books are fragments of correspondence about them between the two men, plus a slightly longer description of *X* by Dreiser that was quoted in his widow's memoir of him.* It seems that in *X* Fort advanced the notion that human civilization is under the remote control of a conspiracy of Martians – or, at least, entities now dwelling on Mars. This alien community deployed rays that were capable, according to Dreiser's description, of *projecting* the entirety of the earth's environment, us included, onto the earth. A rather similar controlling race featured in *Y*, only this time it was hidden at the South Pole; the enigmatic Kaspar Hauser (*c*1812–1833), according to Fort, came from Y-Land and was murdered to stop him telling us about it.

In *Wild Talents* Fort explained himself and his work in his usual inimitable, slightly tortured style:

> I don't do things mildly, and at the same time much enjoy myself in various ways: I act as if trying to make allness out of

* *My Life with Dreiser* (1951) by Helen Dreiser (d1955). Aside from Dreiser, Fort's supporters and admirers during his lifetime included such luminaries as Booth Tarkington (1869–1946), Clarence Darrow (1857–1938) and Oliver Wendell Holmes Jr (1841–1935).

something. A search for the unexplained became an obsession. I undertook the job of going through all scientific periodicals, at least by way of indexes, published in English and French, from the year 1880, available in the libraries of New York and London. As I went along, with my little suspicions in their infancies, new subjects appeared to me – something queer about some hailstorms – the odd and the unexplained in archaeological discoveries, and in Arctic explorations. By the time I got through with the "grand tour," as I called this search of all available periodicals, to distinguish it from special investigations, I was interested in so many subjects that had cropped up later, or that I had missed earlier, that I made the tour all over again – and then again had the same experience, and had to go touring again – and so on – until now it is my recognition that in every field of phenomena – and in later years I have multiplied my subjects by very much shifting to the newspapers – is somewhere the unexplained, or the irreconcilable, or the mysterious – in unformulable motions of all planets, volcanic eruptions, murders, hailstorms, protective colorations of insects, chemical reactions, disappearances of human beings, stars, comets, juries, diseases, cats, lampposts, newly married couples, cathode rays, hoaxes, impostures, wars, births, deaths.

The title *The Book of the Damned* has caused some confusion. To someone who's merely seen the title it might seem the book should be a work of occultism, perhaps a modern version of the *Tibetan Book of the Dead*. In fact what Fort meant by the term "damned" were those "facts" which he believed were being irrationally denounced or ignored by science, which couldn't countenance them in case of making obvious the countless cracks in science's ivory edifice. As he explained:

> By the damned, I mean the excluded.
>
> We shall have a procession of data that Science has excluded.
>
> Battalions of the accursed, captained by pallid data that I have exhumed, will march. You'll read them – or they'll march. Some of them livid and some of them fiery and some of them rotten.
>
> Some of them are corpses, skeletons, mummies, twitching, tottering, animated by companions that have been damned alive. There are giants that will walk by, though sound asleep.

There are things that are theorems and things that are rags: they'll go by like Euclid arm in arm with the spirit of anarchy. Here and there will flit little harlots. Many are clowns. But many are of the highest respectability. Some are assassins. There are pale stenches and gaunt superstitions and mere shadows and lively malices: whims and amiabilities. The naive and the pedantic and the bizarre and the grotesque and the sincere and the insincere, the profound and the puerile.

It never seemed to occur to Fort that his conclusions, as he searched for damned incidents among old newspaper and magazine files at the New York Public Library, were based on an astonishingly dubious premise: that everything you read in the newspapers is true. Bearing in mind that, at the time, the attitude of most of the US press to facts was much like that displayed today by the kind of tabloids you see at the supermarket checkout, this was an even riskier assumption. Especially during the Silly Season when nothing much by way of real news was going on, it was regarded as perfectly acceptable practice, particularly among those provincial newspapers where Fort made some of his most startling discoveries, for reporters simply to make stories up. No wonder, then, that an intelligent (if blinkered) man, confronted by this mess of indiscriminately mixed truth, falsehood, exaggeration and misconception, came to some very strange conclusions.

For example, that the earth does not rotate. Actually, he was prepared to compromise about this: it *might* rotate but, if so, only about once a year. Similarly, it hasn't been proved that the earth is round:

> Shadow of the earth on the moon? No one has ever seen it in its entirety. The earth's shadow is much larger than the moon. If the periphery of the shadow is curved – but the moon convex – a straightedged object will cast a curved shadow upon a surface that is convex. [*The Book of the Damned*]

Then there was the question of the distance of the moon. Fort decided to apply simple mathematics to this problem. He noted a large volcano on earth may be as much as 5km across. Assuming the moon's craters were volcanic (to be fair, so did many people at the time), and ignoring any considerations of gravity, he guessed lunar volcanic craters must be about the same size as those on earth. Reassessing the face of the moon in these terms, he deduced it could be only about 160km in diameter – and thus about 18,500km away. Similarly, he believed the stars to be much nearer to us than astronomers maintained – and that the term "fixed stars" was no misnomer:

> I suspect, myself, that the fixed stars are really fixed, and that the minute motions said to have been detected in them are illusions. I think that the fixed stars are absolutes. Their twinkling is only the interpretation by an intermediatist state of them. . . . I think that Milky Ways, of an inferior, or dynamic, order, have often been seen by astronomers. Of course it may be that the phenomena that we shall now consider are not angels at all. . . . [*The Book of the Damned*]

Having come to the conclusion that the fall of living things from the sky was almost humdrum, Fort turned to reports of the fall of gelatinous substances. Could the underlying explanation for such anomalies be that *the sky itself* was gelatinous?

> We are merging away from carnal to gelatinous substance, and here there is an abundance of instances or reports of instances. These data are so improper they're obscene to the science of today, but we shall see that science, before it became so rigorous, was not so prudish. . . .
> I shall have to accept, myself, that gelatinous substance has often fallen from the sky—
> Or that, far up, or far away, the whole sky is gelatinous?
> That meteors tear through and detach fragments?
> That fragments are brought down by storms?
> That the twinkling of stars is penetration of light through something that quivers?
> I think, myself, that it would be absurd to say that the whole sky is gelatinous: it seems more acceptable that only certain areas are. [*The Book of the Damned*]

In other words, stars are holes in the shell of atmospheric jelly; they twinkle because the jelly trembles. But in *non-*

gelatinous regions of the sky the stars are instead . . . There do come moments when one has to confess defeat.

Fortean enthusiasts are fond of citing in his – and their own – defence the activities of the distinguished French astronomer Camille Flammarion, in particular his book *L'Atmosphère: Météorologie Populaire* (1888). In this Flammarion described various historically recorded falls of red rain. However, there was a major difference between Fort's speculations and Flammarion's. The Frenchman, while accepting (with reservations) the reports as genuine, attempted to offer scientific explanations of each incident. While these explanations may to our educated eyes seem far-fetched or even fanciful, the important point was that they didn't call upon knowledge beyond the ken of mortal beings; they were rooted firmly in what people of Flammarion's day knew, rightly or wrongly, to be possible. And sometimes he was probably correct in his deductions; he attributed a 1744 fall of red rain to red earth being blown off the slopes of a nearby mountain and mixing with moisture in the clouds. Of a 1608 Provence case Flammarion reports that a rationalist on the spot, a certain M. de Peiresc, demonstrated conclusively that little red drops that had appeared here and there were not a Devil-generated spattering of blood from the atmosphere (the popular theory, encouraged by the local priests) but butter-fly excrement – the town had suffered an infestation just beforehand.

By contrast, Fort's theorizing, where it comprised more than just scattershot guesswork, tended to build unsupported speculation upon unsupported speculation, heedless of logic and certainly heedless of the science he scorned. In his biography Damon Knight compiles a list of items that Fort chronicled as having fallen from the skies: alkali, asbestos, ashes, axes, beef, birds, bitumen, blood, brick, butter, carbonate of soda, charcoal, cinders, coal, coffee beans, coke, fibres, fish, flesh, flints, frogs, gelatinous substances, grain, greenstone, hay, ice, insects, iron, larvae, leaves, lizards, manna, nostoc, sand, seeds, silk, snakes, soot, spiderwebs, stones, sulphur, turpentine, turtles, water and worms. In order to explain these precipitations, Fort posited the existence of a sort of Super-Sargasso Sea, a region somewhere above the earth's atmosphere where, according to *The Book of the Damned* (1919), gather

> Derelicts, rubbish, old cargoes from inter-planetary wrecks; things cast out into what is called space by convulsions of other planets, things from the times of the Alexanders, Caesars and Napoleons of Mars and Jupiter and Neptune; things raised by this earth's cyclones: horses and barns and elephants and flies and dodoes, moas, and pterodactyls; leaves from modern trees and leaves from the Carboniferous era – all, however, tending to disintegrate into homogeneous-looking muds or dusts, red or black or yellow – treasure-troves for the palaeon-tologists and for the archaeologists – accumulations of centuries – cyclones of Egypt, Greece and Assyria – fishes dried and hard . . .

Commenting on one of Flammarion's explanations in *L'Atmosphère* concerning an anomalous fall of dead leaves from the heavens, Fort observes in *The Book of the Damned* that these couldn't have been autumn leaves swept up by a cyclone and deposited days later and at a distance because the fall occurred in April. He offers an "inspiration" as a possibility:

> That there may be a nearby world complementary to this world, where autumn occurs at the time that is springtime here.

That there might be a sort of counter-earth nearby was a recurring notion of Fort's. Sometimes he called this planet Genesistrine, because it was there that terrestrial life had originated – thereby solving at a stroke the mystery of the origin of life on earth! Oh, wait a minute . . .

From *The Book of the Damned* again:

> The notion is that there is somewhere aloft a place of origin of life relatively to this earth. Whether it's the planet Genesistrine, or the moon, or a vast amorphous region super-jacent to this earth, or an island in the Super-Sargasso Sea, should perhaps be left to the researches of other super- or extra-geographers. That the first unicellular organisms may have come here from Genesistrine – or that men or anthropo-morphic beings may have come here before amoebae: that,

upon Genesistrine, there may have been an evolution express-
ible in conventional biologic terms, but that evolution upon
this earth has been – like evolution in modern Japan – induced
by external influences; that evolution, as a whole, upon this
earth, has been a process of population by immigration or by
bombardment. Some notes I have upon remains of men and
animals encysted, or covered with clay or stone, as if fired here
as projectiles, I omit now, because it seems best to regard the
whole phenomenon as a tropism – as a geotropism – probably
atavistic, or vestigial, as it were, or something still continuing
long after expiration of necessity; that, once upon a time, all
kinds of things came here from Genesistrine, but that now only
a few kinds of bugs and things, at long intervals, feel the inspiration.

    Not one instance have we of tadpoles that have fallen to
this earth. . . .

In the same book, describing a 1903 fall of red rain, Fort
writes:

    I think, myself, that in 1903, we passed through the remains of
a powdered world – left over from an ancient inter-planetary
dispute, brooding in space like a red resentment ever since. . . .
    To think is to conceive incompletely, because all thought
relates only to the local. We metaphysicians, of course, like to
have the notion that we think of the unthinkable.

Of course, Fort himself very often *did* "think of the unthinkable" – indeed, most of the time he clung to such thinking
and resolutely refused to contemplate the possibility of
rational, "thinkable" explanations for the supposed anomalies he unearthed. Reading through any of Fort's works is
like reading a chronicle of unsolved murders in which the
author insists every killing must have involved the paranormal – that the notion the murderer might simply have got
away with it is boring, closed-minded and stuffy. Well, yes.
But there's often *very good reason* for being boring, closed-
minded and stuffy!

    Towards the end of his life Fort became very ill, perhaps
with leukaemia. Because he didn't trust doctors – scientists!
– he put off seeking medical advice about his poor health,
instead working as hard as he could to finish *Wild Talents*,
which he succeeded in doing; he died in hospital just a few
hours after having seen the book's first advance copies. But

his spirit didn't die with him: there are still plenty of people devoting their energies to the study of so-called Fortean phenomena, and coming to some conclusions that we can regard as nothing short of truly Fortean. Here's his disciple Tiffany Thayer (1902–1959) writing in 1952 (*Doubt*, issue #39):

> I am able to imagine an "Earth" which is a small ball of cosmic-ice, changing ever so slowly to a slightly larger cube of cosmic-ice, then bursting (hatching?) into a ball several times as large as before, scratching whatever of the elements within it are to be scratched. (Luna, I think, is such an ice-ball today, and should be watched for the appearance of corners.)

Probably the foremost current journal of matters Fortean – aside from those supermarket tabloids – is the UK monthly *Fortean Times*, which has been publishing since 1991 and has come to assume a fair amount of cultural respectability. And probably the foremost current researcher into anomalous – i.e., Fortean – phenomena is the US writer William L. Corliss (b1926), author of some 50 books, many of them enormous, with titles like *Strange Phenomena: A Sourcebook of Unusual Natural Phenomena* (1974), *Incredible Life: A Handbook of Biological Mysteries* (1981), *The Unfathomed Mind: A Handbook of Unusual Mental Phenomena* (1982), *Inner Earth: A Search for Anomalies* (1991), *Ancient Infrastructure: Remarkable Roads, Mines, Walls, Mounds, Stone Circles – A Catalog of Archeological Anomalies* (1999) and *Remarkable Luminous Phenomena in Nature: A Catalog of Geophysical Anomalies* (2001). Corliss's industry and dedication are truly astonishing, and it would be a cold heart that failed to admire him, however grudgingly.

It seems almost as if the study of Fortean phenomena can infect otherwise rational people with its own madness. Damon Knight was a science fiction writer and editor of some considerable note,* yet his in many ways excellent

---

* The false stereotype is widely circulated in the mainstream media that sf and fantasy writers are also believers in UFOs, psi and all the rest of it. Yes, obviously, some are. In general, however, bogus science is significantly less likely to be well received in this community than by members of the general public, journalists notably included.

biography of Fort reads in some respects like a diary of Knight's descent into gullibility. Try as one might to persuade oneself that Knight is merely playing with ideas, that he isn't taking this stuff seriously, the conviction slowly grows in the reader that, actually, yes, Knight really *has* begun to fall for the seduction of woo-woo.

The Fortean phenomenon *par excellence* is spontaneous human combustion (SHC), to which numerous books have been devoted – perhaps the most often cited being *Fire from Heaven* (1976) by Michael Harrison (1907–1991). The psychic Sylvia Browne is lucky enough to have her spirit guide Francine to explain to her many mysteries of existence that are beyond the conception of mere mortals, and the mechanism of spontaneous human combustion is one of these. As Browne recounts in *Secrets & Mysteries of the World* (2005):

> [Francine] said that SHC is caused by a buildup of phosphorous [*sic*], which is highly flammable – that's what causes the body to implode upon itself and start burning from the inside out.
>
> I've talked to scientists who allow that the human body is made up of so many minerals and elements that if too much iron is present, for example, we'll die; or if we have too much copper, our liver will fail. So why couldn't we have too much phosphorous [*sic*] (which, by the way, is used in fireworks due to its high combustibility) and just implode?

Francine's expertise extends also to various aspects of cosmology. For example, black holes:

> . . . a black hole is like the universe's vacuum cleaner: A star explodes (or implodes) and creates a type of crater in the atmosphere . . .

Perhaps it might be better to skip black holes and move on to the Big Bang theory:

> That's . . . why the big bang theory is wrong: Like us, the universe has always existed . . . Scientists without a spiritual base . . . had to come up with some plausible explanation like

the big bang theory, but why is it so hard to believe that if God always was, so was all of what God made? Otherwise, God is imperfect – and that's both untrue and illogical.

Returning to more strictly Fortean concerns, Francine has offered an explanation for the Bermuda Triangle. Apparently what causes all the disappearances is that aliens – those same aliens who, arriving on earth, populated Atlantis – long ago constructed there "an intergalactic highway in which people could transport themselves from one planet to another", a device akin to the one featured in the movie *Stargate* (1994) and its TV spinoffs:

> People would place themselves in these tubular chutes to be sent up or down to another planet. [Francine] said that the only problem is that we don't have the technology to understand its concepts, nor do we know how to get in touch with the planet we're trying to reach so that they can employ the mechanism to utilize it. Imagine getting on an elevator without knowing how to push the buttons . . .

Luckless ships and aircraft drift into the transportation system unwittingly and then are whipped away to distant parts of the universe, never to be seen here on earth again. Since the mechanism isn't always functioning, sometimes the Bermuda Triangle is safe; there's no way of telling when these times are, and this very unpredictability is of course an additional factor in the Triangle's danger.

It's not *quite* true to say there's no way of telling if the Triangle's safe or dangerous. It's just a matter of knowing what to look for: "Francine says that they're like envelopes of time that open and close."

Browne – or Francine – also hazards a guess that it might have been the bizarre atmospheric effects he observed when he reached the Bermuda Triangle that so confused Christopher Columbus that he ended up in the West Indies, rather than at his intended destination. I'm not so sure about this. As his intended destination was Asia, it seems more likely that what barred his progress was more likely the Americas, stretching for thousands of kilometres to both north and south.

The "Martian Mermaid" caused
a brief furore in 2008, with
some suggesting the
Copenhagen statue must reflect
a Danish racial memory of
arrival from Mars. Alas, the
object's just a stone

The classic text on the Triangle is of course Charles
Berlitz's *The Bermuda Triangle* (1974). If we're to believe the
account of Berlitz and his many successor Forteans, since
the time of Columbus an astonishing number of ships and
aircraft have inexplicably gone missing in the imaginary
triangle formed by Bermuda, Puerto Rico and the Florida
coast. In *The Bermuda Triangle Mystery – Solved* (1976) Larry
Kusche (b1940), with a true grit that almost defies belief,
painstakingly goes through these disappearances case by
case and demonstrates they're far from as mysterious as
claimed. As a single example, here's a short passage from
the Berlitz book:

> The *Sandra* was a square-cut tramp steamer, decorated here
> and there with rust spots along her 350-foot length. Radio-
> equipped and loaded with 300 tons of insecticide, she leisurely

> The evidence seems to show that sometimes frogs fall
> from above and sometimes they don't.
> – Colin Wilson & Christopher Evans (eds),
> *The Book of Great Mysteries*, 1990

thumped her way south in the heavily traveled coastal shipping lanes of Florida in June 1950.

The crewmen who had finished mess drifted to the aft
deck to smoke and to reflect upon the setting sun and what
the morrow might bring. Through the tropical dusk that
shrouded the peaceful Florida coastline they watched the
friendly blinking beacon at St Augustine. The next morning
all were gone. . . . They had silently vanished during the
night under the starlit sky. . . .

Kusche was able to ascertain, through Lloyd's of London,
that the *Sandra* was not 350ft (106.7m) but 185ft (56.4m)
long – a matter of some importance when considering the
vulnerability of a vessel to harsh seas. He did, however, find
out that the weather in June 1950 was excellent; Berlitz got
that bit right. What Berlitz got wrong, alas, was the date:
again according to Lloyd's, the *Sandra* was passing St
Augustine – if indeed it ever got that far – in early April
1950, not June of that year. And in early April the Atlantic
shipping lanes off the Florida coast were suffering a storm
only just below hurricane strength. Not so surprising, then,
that a small tramp steamer was lost without trace.

And if Berlitz gets such straightforward data wrong,
why believe anything else he says? Would you buy a used car
from someone who habitually misstates the mileage by
50%?

But Kusche, very validly, points out something else. He
did the necessary research to discover the "facts" underlying
Berlitz's account are bunkum; not every reader of Berlitz's
book can be expected to have the time, resources or ability
to do likewise. However, what every critical reader *can* do is
realize, from internal evidence alone, that the passage is
nonsense. Unless Berlitz had actually been aboard the

Sandra, he *cannot* have known all the stuff about going to the rail for a smoke, seeing the St Augustine beacon, etc. No doubt Berlitz would have claimed he was throwing in such invented detail solely in order to provide a little atmosphere. What he was doing in reality, of course, was building an entirely false image in the reader's mind of the weather conditions, thereby creating the semblance of a mystery out of something distinctly non-mysterious.

But here's the problem facing anyone trying to deal with Fortean phenomena. Kusche deserves high credit for his labours in untangling at least part of the web of mythology and mythopoeia surrounding the Bermuda Triangle, and he has done similar sterling work in *The Disappearance of Flight 19* (1980); but how can even someone like Kusche keep up with the indefatigability of a William L. Corliss, who has spent a lifetime cataloguing and cataloguing and cataloguing reports that seemingly describe phenomena that cannot be easily explained? By definition, it takes far less time, effort and resources to catalogue a report that cannot be easily explained than it does to explain it. If Sylvia Browne tells us (as she does) that crop circles are created by visiting extraterrestrials, and that the proof is that her spirit guide Francine has told her so, how can *anyone* undertake sufficient research to demonstrate otherwise? And, really, therein lies the global explanation of Fortean phenomena, the explanation that neatly deals with the vast majority of them and quite probably all: the place to look for the causes of Fortean phenomena is not outwards, to the universe, but inwards, to the credulous human mind.

Some people really believe there were advanced technological civilizations on earth long before the dawn of history, and that it was from these that the earliest known human civilizations sprang. Undoubtedly there are huge gaps in our knowledge of the archaeological past and, while the deductions of science about the contents of those gaps are by and large probably fairly accurate, it must be recognized that they're *deductions*, not proven facts. In August 2008, as reported in *Science* by Charles C. Mann, archaeologists, guided by native people, discovered an extensive network of sophisticated and quite large towns in the Upper Xingu region of western Brazil, for long thought to be virgin tropical rainforest. The civilization responsible seems to have been destroyed by the invading Europeans, both murderously and through disease.

But genuinely interesting discoveries like this tend to be swamped by the gushings of a million and one bogus theoreticians, each clamouring louder than the other. In *Worlds Before Our Own* (1980) Brad Steiger (b1936) cited a US public opinion poll which showed that more people would be interested in the discovery of Atlantis than in the Second Coming.* Dipping one's toe into the atlantological floodwaters one begins to get the feeling that each and every one of the respondents to that opinion poll has published a book on the subject.

---

* Steiger's own theory is that Atlantis is just a distorted memory of a previous civilization that may have been global. Subscribing to the notion that civilization is cyclic – an idea originated by Giambattista Vico (1668–1774), although most of Vico's contemporaries seem to have thought it dubious – and that there have been several cycles before the current one, Steiger supports his hypothesis with accounts of contemporaneous dinosaur and human footprints, consideration of ancient technologies (a mere shadow of *even earlier* glories, natch), and much of the evidence that has prompted other writers to put their faith in ancient astronauts. Some of the evidence, he says, points to the civilization before ours having destroyed itself by nuclear holocaust

The Atlantis legend was born in the 4th century BC in the *Critias* and *Timaeus* of Plato (428–348BC), who claimed therein that his ancestor Solon (c638–558BC) had visited Egypt and there been told by a priest about documentary accounts of a civilization beyond (or at least in the region of) the Pillars of Hercules, which civilization had suffered violent destruction hundreds or thousands (interpretations vary) of years earlier. Often when reading accounts of Plato's popularization of the Atlantis tale one's given the impression that he had it first hand, or nearly so, from Solon. Solon died about 130 years before Plato was born. What Plato knew of Solon was, whether Plato realized it or not (and one suspects he did), a mixture of genuine history, embellished history, and outright legend – fair game, in other words, for Plato to use as basis for some legends of his own.

According to translations of Plato, Atlantis was "larger than Libya and Asia together" – which would seem to make the lost land mighty large! The proportions become more reasonable when one realizes that by "Asia" he meant roughly what we would today call Asia Minor. But there's a further quibble with the translation. The Greek word *meizon* can indeed mean "larger" but it can also mean "mightier", and in Plato's context – his next topic is Atlantis's powerful armies invading the eastern Mediterranean before finally being beaten back by plucky Athens – this latter translation seems the more likely correct one.

Divers atlantological scholars have focused on the fact that in *Timaeus* Plato declared, through his fictionalized character Critias, that the tale of Atlantis was true. The trouble with such contentions is that, throughout his dialogues, Plato frequently claimed as true items that we know neither he nor his contemporaries would believe for one moment: his very purpose in the protestation of veracity was to indicate he was telling a whopper – that what was forthcoming was not a piece of history but a parable.

Even so, it's legitimate to claim that, while the details could have been invented, Plato might have been basing his fiction on a smidgen of genuine truth: there might well have been a great seafaring nation that perished as a consequence of some enormous cataclysm. It's now generally thought by science that the basis of the Atlantis legend could have been the destruction of the Minoan civilization of Crete when the volcanic island of Thera (Santorini) erupted in c1400BC.

This eruption was the second largest in known history, and significantly larger than the 1883 Krakatoa eruption that killed some 40,000 people in Java and Sumatra. Using as a measure the amount of material expelled during the eruption, Krakatoa weighs in at $25km^3$ and Thera, according to research done in 2006, at $60km^3$. (Both are dwarfed by the 1815 eruption of Tambora in Indonesia, at $100km^3$.) But that fails adequately to depict the true scale of the event, which some experts estimate to be around *ten times* as violent as Krakatoa.

The explosion, or the collapse of Thera's giant volcanic cone into the water, generated a tsunami of size similar to the one that devastated much of the Indian Ocean coastline at Christmas 2004; work done in Crete in the 2000s began to reveal evidence of the massive tsunami that ravaged the island, and the flourishing Minoan civilization there. Since the Minoans were maritime traders and had a navy that dominated the Mediterranean, their cities were primarily coastal, increasing the scale of the disaster. The naval and trading fleets must have been crushed, and quite likely destroyed down to the last vessel. Even those Minoan cities further inland, such as the capital, Knossos, would have been subjected to mighty ash falls, extended cooling, sulphuric rain . . .

One might have thought all this dramatic enough for even the most avid of dead-civilization fanciers, but no: it lacks the important element of conspiracy, as in the conspiracy by orthodox archaeology to conceal from public knowledge the truth about our species' history, for reasons of . . . well, the motives are often pretty hard to establish, but they frequently seem to be rooted in a sort of academic protectionism.

Sylvia Browne, in her *Secrets & Mysteries of the World*, displays a certain discomfort with the fact that – while her psychic visions and her spirit guide tells her "the eastern part of Atlantis was off the coast of Spain and Africa, and . . . the western part extended into the Caribbean and the Yucatán Peninsula, also encompassing the Bermuda

Triangle and the Sargasso Sea" – there's quite a lot of physical evidence to suggest the Atlantis legend was born from
the reality of the Thera eruption. So she attempts a compromise:

> Atlantis also had small adjacent islands – of which Santorini was
> one – much like Catalina lies off the coast of California . . .

Perhaps her proofreader pointed out there was a minor
problem with this hypothesis, in that Santorini lies not in
the Atlantic, where Browne fixes the rest of the lost continent, but in the middle of the Mediterranean Sea, because
there's what looks like a hastily added parenthesis:

> . . . (yet Santorini was farther from Atlantis than Catalina is
> from the California coast).

An interesting question rarely posed by those who believe
the account in the *Critias* to be historical is this: how come
all the *other* historians aside from Plato knew nothing of
Atlantis? It's beyond credibility that they simply *missed* it.
Surely the wars with the Atlantean invaders and the eventual foundering of their homeland would have been a major
feature of history and legend all over the region. Herodotus
(*c*484–*c*425BC), who travelled in Egypt and questioned the
very same priestly communities from whom Solon is
supposed to have gained the information, recorded nothing
about Atlantis. And Herodotus, unlike Plato, was what today
we'd call a professional historian: surely he'd have been
onto a nice juicy story like this one like a shot.
    A strong strand of atlantological thought has it that,
before their near-annihilation, the Atlanteans – using that
term to describe a hypothetical ancient supercivilization,
wherever on the globe it might have been centred – passed
on to the more primitive peoples of the world their mathematics, culture, history, legends and science. There's an
enormous glitch right in the middle of such hypotheses. It
would appear the Atlanteans never thought, amid all this
generosity, to pass on knowledge of what one might regard
as the most fundamental of all civilizing skills: agriculture
and animal husbandry. The emergence of agriculture in

Binding embellishment on the
11th edition of Donnelly's classic
Atlantis: The Antediluvian World

different places throughout the world has been fairly well mapped, and it's evident the development occurred at all sorts of different times in different places. This doesn't accord with a single transference of knowledge from the Atlanteans. Further, while obviously the information frequently travelled from one population to the next, it's also evident that very often agriculture was independently invented by a population.

And one can't help wondering what all these ancient high-tech civilizations used for their fuels and raw materials: would they not have severely depleted the world's supplies of coal, oil, gas . . .? Technological civilizations, until they learn better, devour these things, and some – like coal, oil and uranium – aren't replaceable. It's feasible, perhaps, that the Atlanteans made the jump straight from a primitive technology to one that relied solely upon renewable resources, but this doesn't seem to accord with the many accounts of devastating superweapons and the like.

The classic text of the modern age of Atlantean studies is incontestably *Atlantis: The Antediluvian World* (1882) by Ignatius Donnelly (1831–1901). Donnelly, at some very considerable length, described Atlantis as the home of the original Aryans. The Aryans built a major technological civilization, but then either destroyed their homeland with the fearsome weapons their technology had given them or sufficiently annoyed God with their loose living that He arranged the cataclysm that destroyed them.

His introductory chapter is called, plainly enough, "The Purpose of this Book", and begins:

This book is an attempt to demonstrate several distinct and novel propositions. These are:

1. That there once existed in the Atlantic Ocean, opposite the mouth of the Mediterranean Sea, a large island, which was the remnant of an Atlantic continent, and known to the ancient world as Atlantis.

2. That the description of this island given by Plato is not, as has been long supposed, fable, but veritable history.

3. That Atlantis was the region where man first rose from a state of barbarism to civilization.

4. That it became, in the course of ages, a populous and mighty nation, from whose overflowings the shores of the Gulf of Mexico, the Mississippi River, the Amazon, the Pacific coast of South America, the Mediterranean, the west coast of Europe and Africa, the Baltic, the Black Sea, and the Caspian were populated by civilized nations.

5. That it was the true Antediluvian world; the Garden of Eden; the Gardens of the Hesperides; the Elysian Fields; the Gardens of Alcinous; the Mesomphalos; the Olympos; the Asgard of the traditions of the ancient nations; representing a universal memory of a great land, where early mankind dwelt for ages in peace and happiness.

6. That the gods and goddesses of the ancient Greeks, the Phoenicians, the Hindoos, and the Scandinavians were simply the kings, queens, and heroes of Atlantis; and the acts attributed to them in mythology are a confused recollection of real historical events.

7. That the mythology of Egypt and Peru represented the original religion of Atlantis, which was sunworship.

8. That the oldest colony formed by the Atlanteans was probably in Egypt, whose civilization was a reproduction of that of the Atlantic island.

9. That the implements of the "Bronze Age" of Europe were derived from Atlantis. The Atlanteans were also the first manufacturers of iron.

10. That the Phoenician alphabet, parent of all the European alphabets, was derived from an Atlantis alphabet, which was also conveyed from Atlantis to the Mayas of Central America.

11. That Atlantis was the original seat of the Aryan or IndoEuropean family of nations, as well as of the Semitic peoples, and possibly also of the Turanian races.

12. That Atlantis perished in a terrible convulsion of nature, in which the whole island sunk into the ocean, with nearly all its inhabitants.

13. That a few persons escaped in ships and on rafts, and
carried to the nations east and west the tidings of the
appalling catastrophe, which has survived to our own time in
the Flood and Deluge legends of the different nations of the
old and new worlds.

If these propositions can be proved, they will solve many
problems which now perplex mankind; they will confirm in
many respects the statements in the opening chapters of
Genesis; they will widen the area of human history; they will
explain the remarkable resemblances which exist between the
ancient civilizations found upon the opposite shores of the
Atlantic Ocean, in the old and new worlds; and they will aid us
to rehabilitate the fathers of our civilization, our blood, and
our fundamental ideas – the men who lived, loved, and
labored ages before the Aryans descended upon India, or the
Phoenician had settled in Syria, or the Goth had reached the
shores of the Baltic.

The fact that the story of Atlantis was for thousands of
years regarded as a fable proves nothing. There is an unbelief
which grows out of ignorance, as well as a scepticism which is
born of intelligence. The people nearest to the past are not
always those who are best informed concerning the past.

Donnelly dated the Atlantean civilization to the period of
the European Bronze Age and early Iron Age, and was quite
clear it had been as much in communication with the
peoples of the Americas as with those of Europe:

It is probable that the ships of the Atlanteans . . . found the
sea impassable from the masses of volcanic ash and pumice.
They returned terrified to the shores of Europe; and the shock
inflicted by the destruction of the world probably led to one of
those retrograde periods in the history of our race in which
they lost all intercourse with the western continent.

Even so, the Atlanteans did not immediately degenerate
into such lowly brute beings as ourselves. They did things
like build the Great Pyramids as repositories of Atlantean
knowledge in case of future disasters of similar devastating
scale. All in all, Donnelly is quite emphatic,

Modern civilization is Atlantean. Without the thousands of
years of development which were had in Atlantis, modern civilization could not have existed. The inventive faculty of the

present age is taking up the great work of delegated creation
where Atlantis left it thousands of years ago.

Such notions were, predictably, leapt upon eagerly by the
Theosophists, who made of them the usual Theosophist
porridge; and the Aryan connections excited various racists
and Nazis. Among many consequences of the book's publication was that British Prime Minister W.E. Gladstone
(1809–1898), hugely impressed, requested funds from the
Treasury to mount an expedition in search of the lost continent.

Really, though, what Donnelly offered was a civilization
that anyone could use as a canvas onto which they could
paint the perceived virtues of their choice. It's not really
hard to guess the preconceptions of Karl Georg Zschaetzsch
(1870–?) when he speculated about his Aryan ancestors in
Atlantis: Die Urheimat der Arier (1922; "The Cradleland of the
Aryans"): Atlantis was Eden, and in it dwelled the Aryan
master-race, who were vegetarians. The tale of Eve and the
corrupting apple is really one about a wicked female non-
Aryan who began importing cider, so that soon Aryan civilization was reduced to an incompetent alcoholic shambles.
Such disgusting scenes were mercifully ended when the tail

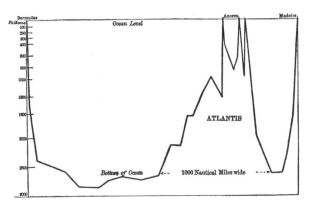

The Atlantic floor, according to Donnelly, with the peaks of Atlantis
today peeping above the waves as the Azores

of a comet ravaged the unhappy, degenerate continent, ripping it apart and plunging the remnants beneath the waves, leaving only three survivors to struggle to the mainland: Wotan, his pregnant sister, and his daughter. The sister died giving birth to a son; luckily a she-wolf was found to suckle the infant. But this could be only a short-term solution to Aryan survival, and incest was out of the question; accordingly the trio began interbreeding with the ghastly aboriginals. This is why the best of us today are so magnificent; but it's also why the vast majority of us debase ourselves by being so repulsive as to eat meat and drink alcohol.

In addition to *Atlantis: The Antediluvian World* Donnelly wrote *Ragnarök: The Age of Fire and Gravel* (1883; sometimes reprinted opportunistically as *The Destruction of Atlantis*) – which prefigured the ideas of Immanuel Velikovsky (see page 77) by ascribing much of the course of prehistory to the earth's having been struck by a comet – and two books seeking to prove the works normally attributed to William Shakespeare (*c*1564–1616) were written in fact by Sir Francis Bacon (1561–1626). The massive *The Great Cryptogram* (1888) set out in its first volume to show all the evidence Donnelly could find or invent in support of his hypothesis; the second volume, which was received with general derision, was a "decoding" of the Shakespearean plays through immensely convoluted cryptography to reveal supposed messages placed there by Bacon. Rather like Michael Drosnin's *The Bible Code* (1997) a century later, this exercise tended to reveal more about the analyst than about the work under analysis. Donnelly's later and lesser *The Cipher in the Plays, and on the Tombstone* (1899) revisited some of this territory while also suggesting that some of the vituperation directed towards *The Great Cryptogram* might have originated with the Rosicrucians, intent on keeping the secret of one of their number, Bacon.

The prime location Donnelly suggested for Atlantis was where the Azores now lie, in the middle of the Atlantic some 1500km west of Portugal, and it's quite often proposed by others that these rocky volcanic outcrops could be the mountaintops of the otherwise sunken Atlantis. Unfortunately, the seabed in which the Azores stand has been underwater for millions of years, putting it well beyond any feasible human timescale.

A main source for Donnelly's researches was the "translation" of the Mayan book known as the Troano Codex done in 1864, complete with much about Atlantis's history, by the Abbé Charles-Étienne Brasseur de Bourbourg (1814–1874). De Bourbourg in turn relied hugely in his researches on the work of the Bishop of Yucatán, Diego de Landa (1524–1579), who demonstrated to his own satisfaction that the undeciphered language used by the Maya had an alphabetical basis (i.e., had letters, like English, rather than ideograms or pictograms) and was even able to match the Mayan letters to their phonetic equivalents in Spanish! Alas, written Mayan is actually a hieroglyphic language; de Landa's work was a massive exercise in self-delusion. De Bourbourg's exercise, based on it, is thus a complete farrago. And, to reiterate, Donnelly made much use of de Bourbourg . . .

Another "translation" of the Troano Codex was done by Augustus Le Plongeon (1826–1908) as *Queen Moo and the Egyptian Sphinx* (1896) and demonstrates that the Maya came originally from ancient Egypt. Le Plongeon's book also revealed the existence in the Pacific of another Atlantis-like continent, with a civilization to match. This was Mu, whose story was hugely developed by Colonel James Churchward (1852–1936) in *The Lost Continent of Mu* (1926) and its successors, culminating in *Cosmic Forces of Mu* (1934). Indian or Tibetan monks apparently gave Churchward a number of tablets in a language that he was told was Naacal, the very first written language of all. Only two mystics in the world could understand Naacal, but Churchward was lucky enough that they taught him, too, how to read it. Thus educated, he was able to translate his tablets to reveal that Mu sank into the Pacific 12,000 years ago; only a few of its 64 million inhabitants survived, and these fortunates populated various Pacific islands. Their descendants formed modern *Homo sapiens*, of which species only members of the Aryan race approach the lofty virtue of their Muan ancestors; the rest of us are descended from products of interbreeding between Muans and the hideous primitives of those islands.

John Dee (1527–*c*1608) called North America Atlantis, and there was a school of thought that the Americas had indeed been the Atlantis of legend. One difficulty with this hypothesis is that the Atlanteans were supposed to have waged successful maritime campaigns against the great powers of the Mediterranean. It's hard to imagine how, without modern technology, Atlantis could provision or communicate with a fleet operating in uniformly hostile territory several thousand kilometres from home. Even navigating accurately from the Americas to the Mediterranean represented an enormous challenge. Of course, it could be argued that the Atlanteans did have a superior technology, but in that case (a) we'd be finding its relics all over the place, mixed in with the other ancient-world military relics we do discover, and (b) the Atlanteans would – no matter the pluck of the Athenians *et al.* – have won the wars so comprehensively that all the countries of the Mediterranean would thereafter have been Atlantean colonies.

One of the more unusual suggestions about Atlantis's location came in *The Riddle of Prehistoric Britain* (1946) by the Fleet Street journalist William Comyns Beaumont (1873–1956), who thought Britain was the home not only of the long-lost Atlantean technological supercivilization, complete with its aircraft and its battle weapons, but also of many of the civilizations in history that we've always regarded as being sited in other parts of the world. Scotland was the particular beneficiary of Beaumont's historical generosity, being home to ancient Greece, Babylonia, Egypt and Israel. He identified Dumbarton as the original Athens, while Ben Nevis, in Scotland's northeast, was the original Olympus. The monument of Avebury, whence Abraham led the Chosen People from their oppression in the Orkney Isles, could well have been Thebes, or perhaps it was the dragon teeth Cadmus sowed . . . or perhaps it was *both*, because Beaumont ran into a certain amount of difficulty finding sufficient geographical landmarks in Britain to correspond to all the places mentioned in ancient history and had to advance the ancillary hypothesis that different civilizations in their turn renamed all the settlements, mountains, rivers, etc.

Eventually a double comet fell from the sky, landing near Jerusalem/Edinburgh and knocking the earth into an orbit further from the sun. (The cold, damp climes that we associate with Britain today are a result of this increased orbital distance; when Britain was Atlantis it enjoyed a far balmier climate.) Atlantis's surviving inhabitants emigrated to other parts of the world where – and here's the cunning part of Beaumont's scheme – they *gave their settlements the names of places back in the Old Country*, just as the European settlers did in North America. So the reason we think that, say, Egypt was in northern Africa is that the people who arrived there called this new territory after their old stamping ground of Egypt, Britain.

Naturally enough, Atlantis/Britain was also the Holy Land, with the real Jerusalem being what we now call Edinburgh. Most of the Old Testament happened while Britain was still Atlantis, before the comet struck, but after the catastrophe this was where the events of the New Testament were played out – with Somerset being known as Galilee, Bristol as Sodom, London as Damascus, etc. Glastonbury, which had earlier been the Garden of Eden, was now the birthplace of Christ; much later Joseph of Arimathea famously sailed here from Edinburgh. The chronology is difficult to juggle, but it must have been after the time of Christ that the Romans destroyed Jerusalem/Edinburgh as part of their invasion of Britain, which began – according to orthodox history books – in 44BC.

The Riddle of Prehistoric Britain was followed by *Britain: The Key to World History* (1947), which continued in similar vein: "Above all, the history of the Old Testament is the history of Atlantis."* In this book Beaumont seems to link the destruction of Atlantis to the belief by geologists that around the late Tertiary/early Quaternary boundary, perhaps four million years ago, land in the triangle formed by Norway, Iceland and Scotland collapsed into the sea, as did further lands between England's south coast and Brittany; the estuary of the Rhine expanded hugely to form

---

* There was also a third, as yet surprisingly unpublished volume, *After Atlantis*.

the North Sea. Certainly those geographical changes did occur. They did not, however, occur as a single catastrophic event, as Beaumont seemed to think; the lands sank not overnight but as part of an ongoing and primarily gradual process of modification of the earth's face.

In the summer of 1998 a Russian team searched for Atlantis on the Little Sole Bank, a subterranean hill on the Atlantic seabed some 150km off the southwestern tip of England. So maybe Beaumont will have the last laugh.

The above were not his first works. Earlier had come *The Riddle of the Earth* (1925) and *The Mysterious Comet, or The Origin, Building-Up and Destruction of Worlds, by Means of Cometary Contacts* (1932). The first of these was in a way ahead of its time, in that it challenged the then-prevalent orthodoxy of uniformitarianism in the earth's geological history by claiming that major changes had come about owing to cometary, asteroidal and meteoric impacts. We now know that this is in fact the case – for example, it seems certain that the catastrophe which, among much else, eventually wiped out the dinosaurs was a cometary impact. In the second book, which saw a shift of publisher from the stuffily scientific house Chapman & Hall to the mysticism-oriented firm Rider, Beaumont took this thesis a long step further, producing what was so much a precursor of the ideas of Immanuel Velikovsky that several supporters of the latter have produced laboured explanations as to why their hero could not possibly have been guilty of plagiarism.

Alfred de Grazia (1919–2006), for example, in his memoir of matters Velikovskian, *Cosmic Heretics: A Personal History of Attempts to Establish and Resist Theories of Quantavolution and Catastrophe in the Natural and Human Sciences* (1984), lists 25 of Beaumont's core notions and has the honesty to admit the exceptional similarities. He adds at the end: "One significant thesis that V[elikovsky] could not have got from Beaumont was that the disturbing comet was Venus, although both identified Quetzalcoatl with the comet." However, he later decides that the coincidence of ideas must be a demonstration that both men had arrived independently at the same conclusions, thereby adding to the credibility of those conclusions, because, after all, how could Velikovsky, working in New York City, possibly have obtained access to these obscure, long out of print British volumes?

The answer to this riddle, the UK social anthropologist Benny J. Peiser pointed out in a 1996 paper in *Chronology & Catastrophism Review*, could well lie in the fact that Beaumont's books were on the shelves of the New York Public Library, where Velikovsky did much of his research.

Beaumont's later book of note was the self-published *The Private Life of the Virgin Queen* (1947), in which he promoted the theory – not original to him – that Francis Bacon was the bastard son of Elizabeth I. Beaumont was firmly convinced Bacon was the true author of the works generally attributed to Shakespeare and had inserted this information, in cipher form, into the various plays and poems. But that, as they say, is a whole 'nother can of bacon.

Beaumont's Atlantean hypothesis – although not his reconstruction and relocation of Biblical events and lands – has had echoes as recently as 2003 in Paul Dunbavin's *Atlantis of the West: The Case for Britain's Drowned Megalithic Civilization*. Atlantis, in this scheme, was a large island in the Irish Sea, and was the pinnacle of a Neolithic civilization that encompassed much of western Europe and was inundated *c*3100BC when the world's rotational axis shifted as a consequence of a cometary collision. As noted above, a cometary collision is *per se* far from an impossible notion, although it's strange there's so little evidence for this one; but the phenomenon of Pole Shift upon which Dunbavin so freely relies is 'way out on the far shores of impossible physics. A cometary impact capable of significantly altering the direction of the earth's axial rotation would have to be of such a magnitude as to blast the planet to bits.

In *The Flood from Heaven: Deciphering the Atlantis Legend* (1992) Eberhard Zangger proposed that the legend could be the Egyptian version of the history of the Trojan War. This naturally differed from Homer's, and would have differed even more by the time it had passed through several hands from Egyptian hieroglyph to Plato. If we accept Plato's statement that the tale was originally told to the visiting Solon in Saïs, capital of Egypt, we can visualize a scenario whereby Solon, not realizing this was the Trojan War he'd been told about, started writing down the tale but for some reason never completed the manuscript. The unfinished work passed down through the family to reach Plato, who was only too keen to recycle it. He began incorporating the tale into his two Dialogues but suddenly, midway through *Critias*, recognized he was writing not about some forgotten island civilization but about Troy, and abandoned the plan. Zangger's interpretation depends upon the assumption – now widely held – that Troy was not so much a city as a nation; his hypothesis is appealing on archaeological grounds, although obviously he's speculating when it comes to Plato's actions.

Or was Atlantis in Sweden? This was the theory of Olof Rudbeck (1630–1702), a professor at Uppsala University and a Renaissance man *par excellence*: in 1650 he was (contestably) the first to discover the lymphatic system and its function, and he was also a distinguished botanist,* architect, shipbuilder, artist, cartographer, composer, musician and singer. His interests in the ancient world and in linguistics began in the late 1660s when his friend and Uppsala University colleague Olaus Verelius (1618–1682) asked him to draw the maps for a published edition of a hitherto unpublished Icelandic saga, the *Hervararsaga*, one of a number brought to Uppsala from Iceland by the reprobate Icelandic scholar Jonas Rugman (1636–1679). The more he read of the saga while preparing its maps, the more Rudbeck became convinced it must have been in Sweden that civilization began, spreading thence to the rest of the world.

Rudbeck threw himself into archaeological and linguistic researches designed to prove his case. It's a sign of the man's genius that a mere by-product of his growing obsession was his invention of a means of approximately dating archaeological remains through study of the layers of soil above them. Using a calibrated rod to measure the soils covering, for example, the ancient burial mounds near Uppsala, he was able to satisfy himself that civilization had

---

* The plant genus *Rudbeckia* was named in honour of Olof and his son, another Olof (1660–1740).

Portrait by Martin Mijtens of Olof Rudbeck

flourished in Sweden a millennium before the Trojan Wars. Was this, perhaps, where the descendants of Noah had set about repopulating the earth after the Deluge? It seemed a reasonable possibility, bearing in mind the floodwaters must have destroyed all other forms of food save fish, and where else in the world would you go for the best fish except Scandinavia?

At first Rudbeck was content with the idea that Sweden must have been the Hyperborea of Greek legend – this is by no means a ridiculous idea. Further detective work enabled him to offer a rereading of the legend of Jason whereby the Argonauts, after seizing the Golden Fleece, included Sweden on their itinerary during their long journey home. Although this involved the Argonauts navigating a complicated route that started up the Don from the Black Sea and through Russia's river network to the Baltic, a route that included a major cross-land trek when they'd have had to drag their ship on rollers, it was not entirely impossible.

But Jason wasn't the only adventurer whose travels were recorded in Classical myth and legend. Lots of the heroes and their companions had made the trip to Hades and back. Might it be possible that Hades wasn't really the land of the dead but somewhere else, perfectly mundane, that was characterized by cold and damp and long periods of darkness? In other words, the Arctic regions of northern Europe. Hello, once more, Sweden.

So far, although Rudbeck was mixing in a very generous ratio of outright speculation with his more grounded deductions, his work was not too overtly fantastical. His next fixation, though, gently and then more swiftly raised him above the surface of reality's plains. Of course, if one accepts that there was a historical Atlantis, then the idea that it could have been in the vicinity of Sweden is no more ridiculous than any other, and certainly some of the evidence Rudbeck produced for his hypothesis of a Scandinavian Atlantis, with its capital at Old Uppsala – nice and convenient for his archaeological researches! – is even today oddly persuasive. For example, one of the Greek words for "island" could also mean "peninsula", so Scandinavia was as well qualified as any of the island possibilities. Perhaps most haunting of all is the reference in Tacitus to there being tell of a second Pillars of Hercules far to the north: what better description could one find of the Øresund, the narrow, treacherous sea passage between southwestern Sweden and the Danish island of Zealand?

All this and much, much more Rudbeck poured into his epic book *Atlantica*, the first mammoth volume of which appeared in 1679, with further colossal volumes appearing periodically until Rudbeck's death in 1702. Later historians would describe *Atlantica* as perhaps the most enormous testimonial to the temptations of erroneous thinking ever produced, but at the time – despite some detractors – it was accorded the same sort of respect, even adulation, as Newton's *Principia* (1687) a few years later. On the strength of it Rudbeck was proposed, even though a foreigner, to the Royal Society. The book might have retained its status as a pinnacle of European scientific endeavour for far longer had it not been that Sweden very soon suffered its own loss of status, stumbling from major European player to conquered nation. In that context, the notion of Sweden as cradle of civilization and ancient superpower seemed – even though this conclusion is in itself illogical – too ludicrous to maintain.*

Centuries later, such is the apparent disloyalty among Swedish academics, Ulf Erlingsson of Rudbeck's own Uppsala University selected Ireland as his candidate for Atlantis in *Atlantis from a Geographer's Perspective: Mapping the Fairy Land* (2004). Since Ireland is very obviously unsunken, the legend of Atlantis must be conflationary, confusing the Atlantean/Irish civilization with the sinking of a different island. This, Erlingsson proposes, was an island inundated in the Dogger Bank region in about 6100BC, when the ancient Lake Agassiz drained into the world ocean, raising sea levels globally. Certainly some prehistoric remains have been discovered in this region, so the notion's not entirely fanciful.

In similar vein was the proposal by Robert J. Scrutton in *The Other Atlantis* (1977) and *Secrets of Lost Atland* (1978) that there coexisted with Atlantis another Atlantis with a confusingly similar name. This other blessed isle, Atland, lay between Britain and Greenland, and survived until 2193BC, when it was abruptly destroyed either by asteroidal impact or as part of the general devastation caused by the comet that, in the works of Velikovsky, made life tough for our ancestors as it looped repeatedly round the earth. Scrutton's hypothesis seems to depend on the Oera Linda Book, even though that's now generally believed to be a 19th-century forgery.

Half a century earlier the Scottish mythologist Lewis Spence (1874–1955) was another to posit a lost Atlantic island civilization other than Atlantis. Of his half-dozen books on Atlantis the most important are arguably *The Problem of Atlantis* (1924) and *The History of Atlantis* (1927), in which he describes Atlantis's sister-continent, Antillia. Spence's Atlantis was about the size of Western Europe and

---

* An enormously readable book on Rudbeck and his obsession(s) is David King's *Finding Atlantis* (2005).

Spence's proposed locations for
Atlantis (A) and Antillia (B)

lay fairly closely off the coasts of northwest Africa and Spain; Antillia, a little smaller, lay between Atlantis and the Caribbean. Communication between Antillia and Atlantis was facilitated by an intervening archipelago. According to Spence,

[T]hese two island-continents and the connecting chain of islands persisted until Pleistocene times, at which epoch (about 25,000 years ago, or the beginning of the Post-Glacial epoch) Atlantis appears to have experienced further disintegration. Final disaster appears to have overtaken Atlantis about 10,000BC. Antillia, on the other hand, appears to have survived until a much more recent period, and still persists fragmentarily in the Antillia group or West India Islands.

Refugees from Atlantis arrived in Europe and settled from about 25,000BC onwards, exterminating the less developed human species, Neanderthal Man, that they discovered there; we know the Atlanteans from fossil evidence as Cro-Magnon Man. Later Atlantean emigrants founded the advanced civilizations of Crete and Egypt. As Antillia foundered, refugees fled to South America to become the Maya.

It was the contention of Charles Berlitz (1914–2003) in *Atlantis – The Eighth Continent* (1984) that we've got past the time when the only way to learn about Atlantis was through study of ancient texts, linguistic similarities or global distributions of plant and animal life: we can now directly examine the remains of lost lands beneath the Atlantic – and beneath 12,000 years' worth of accumulated sludge. Even so, he adduces plenty of circumstantial evidence, such as that eels travel to the Sargasso Sea by way of an underwater river which passes close to where Berlitz thinks Atlantis was, that lemmings indulge in mass suicides by jumping into the sea because they're trying to migrate across what was once dry land to Atlantis,* that the Basques believe themselves descended from the inhabitants of a drowned continent called Atlaintika, and that there are indigenous lifeforms on mid-Atlantic islands (like dogs on the Canaries) that couldn't have got there any other way except from a foundering Atlantis, because the human inhabitants didn't have boats. This last argument seems peculiarly specious since, if we accept Berlitz's line of reasoning, the real puzzle would be how the *people* got there, not the dogs. Of course, that's not a puzzle at all, so the dog mystery collapses.

Most islands of any size in the Mediterranean have been proposed at one time or another as the site of Atlantis. Perhaps the most enterprising endeavour along these lines is the Maltadiscovery Prehistory Foundation, founded and run by the German entrepreneurial archaeologists Dagmar Claire and Hubert Zeitlmair. The MPF's website is full of information about the astonishing archaeological discoveries of this pair, who are also keen to find customers for their holidays and guided tours:

> If you are searching for greater personal meaning and feel that you are being called back home to this primeval land [i.e., Atlantis], we invite you to join us in Malta and experience with us the ancient wisdom of Atlantis which will open you to higher dimensions of awareness and allow you to experience a connectedness with nature as you have never experienced before.
>   The high powered magnetic energy of the Sacred mount of Malta also enhances cellular activity which in turn is conducive to personal transformation through multi dimensional experiences. We seek to integrate the physical and spiritual bodies in order to achieve wholeness of body, mind and soul which in turn leads to good Health.

Golly. In fact Claire and Zeitlmair have a surprising number of precursors in claiming Atlantis for Malta. In 2000 the

---

* An idea earlier voiced by Philip Lutley Sclater – see page 141.

Prehistoric Society of Malta published *Malta: Echoes of Plato's Island* by Anton Mifsud, Simon Mifsud, Chris Agius Sultana and Charles Savona Ventura; since the book's only 87 pages long, that's an impressive author/page ratio. Other Mifsud literary enterprises include *Dossier Malta: Evidence for the Magdelenian* (1997) by Anton and Simon and *Facets of Maltese Prehistory* (1999) edited by Anton with Ventura.

In 2002 the Maltese journalist Francis Galea (b1953) published *Malta fdal Atlantis* ("Malta, Remains of Atlantis") as a result of his investigations of Maltese temples and other archaeological relics. Long before any of them, however, the celebrated Maltese architect Giorgio Grognet de Vassé (1774–1862) published *Compendio ossia Epilogo Anticipato . . . della Famosa Sommersa Isola Atlantide* (1854); this was his own abridgement of a rigorous two-volume dissertation on the hypothesis that's still preserved in the National Library of Malta. He was probably also the author, in 1832 or 1833, of a 19-page essay on the subject bound in as an incongruous supplement to the reprint of Malta's first tourist guidebook.

It is the Mifsuds' contention that the temple-building culture on Malta, wiped out by tsunami around 2200BC, dates back to an initial human inhabitation of the island some 15–18,000 years ago, in the Palaeolithic. Orthodox archaeology maintains that humans didn't arrive in Malta until much later, about 6000 years ago, during the Neolithic. In order to decide the issue, much depends on the age of a pair of human teeth found in a natural cave on the island, and now held by Malta's Museum of Archaeology. Here conspiracy theory enters the picture. During the 1950s and 1960s the Museum had the teeth dated by experts at London's Natural History Museum; the reports quite clearly indicate a Neolithic date for the teeth. However, the Misfuds claim the authorities at the Museum of Archaeology *forged the relevant figures in one of the reports*! Alas for their case, the record preserved at the Natural History Museum tallies with the copy of the report held in Malta.*

---

* The Museum of Archaeology refuses to submit the teeth for analysis using the more recently developed technique of radiocarbon dating, so

An alternative Maltese hypothesis was presented by the Italian journalist Sergio Frau in *Le Colonne d'Ercole – Un'inchiesta* (2002; "The Pillars of Hercules – An Investigation"). In Frau's scenario it was the Greek scientist Eratosthenes (*c*276–*c*195BC) who relocated the Pillars of Hercules to the Strait of Gibraltar; before this the term generally referred to the strait between Sicily and Tunisia. In that case the obvious candidate for Atlantis would be the island of Sardinia, and sure enough Sardinia did have a relatively advanced early civilization about which not much was known: these people were the builders of the 8000 or so megalithic monuments known as nuraghes that adorn the island, and were renowned in the Mediterranean for their bronzework. The earliest nuraghes to have been dated go back to 3500BC, although most are somewhat more recent. According to Frau's hypothesis, a tsunami more or less wiped out the nuragic civilization of Sardinia in about 1175BC. This would be a plausible match to Plato's account, according to some scholars.

And then there's Sicily itself. The Sea Peoples were a culture about whom little is known except that they perpetrated raids on Egypt, Cyprus, Hatti and the Levant during the 12th century BC. The German researcher Thorwald Franke* believes they were the Sicels of Sicily, that Sicily was the island and the Sicels the culture to which Solon's Egyptian priest was referring when he described Atlantis, and that the Atlantean King Atlas can be identified with the Sicilian King Italos, as mentioned by Thucydides (*c*460–*c*400BC) in his history of the Peloponnesian war. Franke believes that when Solon thought the Egyptian priest was referring to the Pillars of Hercules the man really meant the Strait of Messina, the narrow body of water between Sicily and Italy: the Egyptian could well have used a term like "a strait to the west of here" to mean the Strait of Messina whereas a Greek would have assumed the same phrase meant the strait at the mouth of the Mediterranean. As support for his contention, Franke points out that Plato's

one can at least understand where the Mifsuds' conspiracy theory is coming from.

* Who maintains a relevant website at www.atlantis-scout.de.

account describes the Atlanteans having conquered Italy
and North Africa before moving on to invade the region
"within [i.e., east of] the strait"; this would imply the strait
in question had to be east of Italy and North Africa, which
Gibraltar manifestly isn't.

While Sicily, like other islands we've been discussing,
hasn't sunk beneath the waves, it's in a volcanic region and
has suffered accordingly from eruptions, earthquakes and
tsunami; a 1908 quake destroyed the city of Messina with a
death toll of some 60,000. A similar geological catastrophe
might have devastated to the point of extinction the prehistoric civilization of the Sicels. Yet the Egyptian priest,
according to Plato, was insistent Atlantis was entirely
destroyed. Might this not be, Franke posits, merely a reflection of the Egyptian belief in the deity of the Pharaoh: if the
Pharaoh-god were to deal with the problem of the maraud-
ing Sea Peoples, a task that in mundane terms Egypt was
militarily ill equipped to tackle, the way to do it would be by
divine smiting, and, well, when gods smite islands, islands
know they bin smited – not a half-measure in sight. The
priest thus reported the complete, divinely perpetrated
annihilation that *must* have happened rather than the debil-
itating natural catastrophe that actually did.

All of this is feasible, of course; but one feels there's
rather too much special pleading going on.

Growing in popularity as a site for Atlantis is the
submerged island of Spartel, which lies on the Atlantic side
of the Strait of Gibraltar, in the Gulf of Cadiz, just neatly in
the region where Plato said Atlantis was. Although the
island is currently below the waves, research done on sea
levels prevalent as the last glaciation ended shows – according to geologist Jacques Collina-Girard of the University of
the Mediterranean, Aix-en-Provence, in 2001 – that about
19,000 years ago the water level between Spain and
Morocco was 130m below what it is now. Assuming the level
steadily rose, Spartel would have sunk below the waves of its
own accord some 11,000 years ago, a date that tallies well
with some interpretations of Plato's account.

But that's assuming the water level has risen steadily.
More recent research by Marc-André Gutscher of the
University of Western Brittany, reported in the journal

Geology in 2005, suggests Spartel may have been drowned by a tsunami a thousand years earlier than Collina-Girard's estimate: Gutscher found a thick turbidite deposit on the local seabed that indicated the sediments in the area had been subjected to a considerable upheaval around 12,000 years ago. This would still accord reasonably with Plato, and would match his assertion that the downfall of Atlantis was wrought in a single day. A quake the size of the one that devastated Lisbon in 1755 would have been sufficient to do this damage to the region.

Of less cheer to supporters of the Spartel/Atlantis hypothesis was that, according to Gutscher's mapping, the submerged island was quite a lot smaller than had been thought, and there were no signs of any manmade structures.

The Sea Peoples, mentioned above, quite likely came from Spain and have been tentatively identified with an Iron Age people of Andalusia known as the Tartessos culture. As early as 1673, in *Aparato a la Muonarchia Antigua de las Españas en los tres Tiempos del Mundo*, the Spanish author Joseph Pellicer de Ossau Salas y Tovar (1602–1679) proposed that the capital of Atlantis was in the middle of what are today the Doñana marshes, and this idea was reinforced by such 20th-century scholars as Adolf Schülten (1870–1960), in his *Tartessos* (1924).

In the 1990s the German atlantologist Werner Wickboldt (b1943) drew attention once again to the Marisma de Hinojos region of the Doñana National Park in Andalusia, where satellite photos reveal concentric circles of alternating land and water much as Plato described in the capital city of Atlantis. In June 2004 the German physicist Rainer W. Kühne (b1970) published in the journal *Antiquity* a paper called "A Location for 'Atlantis'" that heartily endorsed Wickboldt's suggestion about both the Sea Peoples and the site:

> In fact, near Cadiz there is a rectangular, smooth and even plain which lies on the south coast at the mouth of the Guadalquivir River. It is the plain southwest of Seville through which the Guadalquivir river flows, and where the town of Tartessos was thought to have been located. [Various writers of the 1920s and 1930s] have . . . supposed that this was a possi-

ble location for Plato's Atlantis. In this respect, it is not without interest that large structures have been identified from
recent satellite photos in this part of the lower Guadalquivir
basin. One shows a rectangular structure with a length of
230 metres and a width of 140 metres. It could be a remnant
of a temple of Poseidon, such as that whose length was one
stade (185 metres) and whose width was three plethra (92
metres) . . . A further "quadratic" structure of size 280 metres
times 240 metres could equate to the temple of Cleito and
Poseidon . . .

Other geographical features of the region seem to match
up, according to Kühne, with Plato's description: the plain
of Atlantis could well be the plain reaching from Spain's
south coast to Seville; Atlantis's high mountains could be
the Sierra Nevada and Sierra Morena; and so forth. It is not
at all impossible that the Ancient Egyptians might have
misidentified Spain as an island.

Archaeologists have been slow to embrace Kühne's
hypothesis. One great difficulty in settling the matter is that
the Doñana National Park is a conservation area, so one
can't just go digging big holes in it.

Spain drew attention a little earlier, in the mid-1980s,
when the scholar Jorge Maria Ribero-Meneses proposed
that Atlantis could have been what is now the Danois bank,
a submerged plateau measuring about 50km by 18km and
some 65km off the coast of northern Spain. According to
Ribero-Meneses, tectonic changes at the end of the last
glaciation, some 12,000 years ago, caused a large chunk of
the continental crust to break off here, with the customary
tsunami, huge loss of life, and near-annihilation of civilized
behaviour.

In 2004 a US team led by the architect Robert Sarmast
of First Source Enterprises used sonar to map the
Mediterranean seabed between Cyprus and Syria. Sarmast's
hypothesis, as reported in *Discovery of Atlantis: The Startling
Case for the Island of Cyprus* (2004), is that Cyprus was once a
much larger island, and was connected via landbridge to the
eastern shore of the Mediterranean; Atlantis, which was also
the Garden of Eden, was located on a part of Cyprus now
submerged. His expeditions of 2004 and 2006 to a submarine rise he had already decided must be Atlantis's

"Acropolis Hill" revealed massive structures whose geometrical shapes suggested them to be manmade, such as two straight walls 2km long. Speaking to the BBC he said:

> The hill, as a whole, basically looks like a walled, hillside territory, and this hillside territory matches Plato's description of the Acropolis Hill with perfect precision. Even the dimensions are exactly perfect, so if all these things are coincidental . . . we have the world's greatest coincidence going on.

Sarmast's work and hypotheses were featured alongside others in a SciFi Channel documentary called *Quest for Atlantis: Startling New Secrets* (2006). More significantly, in early 2007 they were the subject of major examination in *Atlantis: New Revelations*, a season premiere of the History Channel series *Digging for the Truth*. The conclusion here was that Sarmast's supposedly artificial features were more likely natural formations. As Sarmast's defenders were quick to point out, the history of early civilization is surely one of taking advantage of natural formations, so what his sonar revealed may well be geomorphological features modified by the ancients. One suspects this is a discussion that's set to last a while.

There was great excitement in early 2009 when the extension of Google Earth to the oceans revealed what appeared to be a network of straight lines – a grid of streets? – on the ocean floor about 960km west of the Canary Islands. They were spotted by UK aeronautical engineer Bernie Bamford when he was playing with the online facility, and he breathlessly informed the UK tabloid *The Sun*. In fact, the lines were artefacts of the data-collection process: the boats bearing the sonar equipment work backwards and forwards following a grid system, and there's a dropoff of data between the lines they sail.*

---

* The effect that Bamford had discovered bore analogies with one that decades earlier excited hollow-earth enthusiasts. NASA released a collage of satellite photos showing the north polar region, and left blank a central area which had yet to be covered by any of the orbiting cameras. This blank was immediately seized upon by people like Brinsley le Poer Trench (1911–1995) as "evidence" of a polar hole leading into the hollow earth.

Bolivia is another proposed locale for the lost continent, and it is to there that UK teams led by John Blashford-Snell started going annually in search of it in the summer of 1998 under the auspices of the Scientific Research Society. In August 2002 an item by Alexei Vranich in the magazine *Archaeology*, published by the Archaeological Institute of America, painted a less rosy picture of the scholarly integrity of these expeditions than the organizers might have wished:

> Controversy erupted last summer when the group announced its discovery of the "lost city" of the Inka [*sic*] . . . in fact the expedition had only reidentified one of the many Precolumbian sites recorded in the 1950s by Hans Ertl, formerly Hitler's photographer. Retracing Ertl's survey, the group used dynamite to clear a six-mile path through the threatened subtropic forest of the eastern Andes.
>
> A protest by local archaeologists was quickly quashed by the expedition's powerful allies, including the unpopular Bolivian president Hugo Banzer, [who had been] made the godfather of the enterprise by expedition leader Colonel John Blashford-Snell. . . .

Vranich laments the fact that the expeditions, because they claim to be searching for romantic places like Atlantis and El Dorado, attract very hefty funding from major corporations – American Airlines, Nikon, Suzuki, etc. – and in addition charge sizeable fees for the participants. Meanwhile, the archaeologists pressed into the expeditions' service by the Bolivian state receive merely subsistence. And he derides one of the Scientific Research Society's other claims:

> To make the case for Precolumbian transoceanic contact, the British-based expedition tried to navigate reed boats from Lake Titicaca to Africa, a feat that its website called "a smashing success." They did navigate the Amazon to the coast of Brazil, but their boats were motor-powered and carried part of the way in trucks.

That buzzing sound you hear is Thor Heyerdahl spinning in his grave.

Perhaps an obvious locale for Atlantis is Antarctica, and

the notion was made to seem even more so by the publicity given to the celebrated Piri Re'is map, dated 1513 and discovered in 1929 in Istanbul. At first sight this appears to show the coastline of Antarctica as it would be if unmasked by snow and ice. The map was discussed at great length by Charles Hapgood (1904–1982) in his *Maps of the Ancient Sea Kings* (1966); he suggested there might have been a very early maritime culture whose boats occasionally sailed along the Antarctic coast at a time when it was free from ice. Hapgood also – earlier, in *Earth's Shifting Crust* (1958) – came up with the Crustal Displacement Theory, an hypothesis which seems, so far as I can figure it, untroubled by basic geology. The notion is that, in addition to the customary plate-tectonic movements of the earth's outer layer, bits of the crust can independently and fairly rapidly slip over the molten mantle by distances up to about 2000km. Thus what might once have been a mid-Atlantic continent could now be straddling the south pole. If one tries to maintain that this happened in historic times or relatively recent prehistory – 10,000 years ago, perhaps – one faces the problem that the sudden appearance of a large landmass at one of the planet's poles would have considerably affected global climate. Further, a vast amount of scientific evidence, drawing upon a number of different disciplines, indicates quite clearly that at no time during the past 400,000 years has Antarctica been other than ice-covered.

Rand Flem-Ath, in books like *When the Sky Fell* (1995; with Rose Flem-Ath) and *The Atlantis Blueprint: Unlocking the Ancient Mysteries of a Long-Lost Civilization* (2001; with Colin Wilson), has enthusiastically promoted the notion that Atlantis was located in what is now western Antarctica. Today largely promoting his hypotheses through his website, www.flem-ath.com, Flem-Ath appears to be relying less on the Crustal Displacement Theory, more on the graphic device that, if we look at the map of the world from above the south pole, with Antarctica at the centre, rather than from above the equator as we usually do, we see quite clearly that Antarctica lies in the middle of the world ocean with a solid mass of other continent all around it, and is indeed beyond the Pillars of Hercules.

The Crustal Displacement Theory, and the notion that

Rand Flem-Ath's depiction of Atlantis/Antarctica as at the
centre of the world ocean

Atlantis lay in Antarctica, were seized with zeal by the UK
unorthodox archaeologist Graham Hancock (b1950),
author of a string of bestselling books. Not since Erich von
Däniken has any writer aroused such archaeological ire. In
the second of these books, *Fingerprints of the Gods: A Quest for
the Beginning and the End* (1995), Hancock offers the thesis
that the ancient and technologically advanced civilization of
Atlantis died out soon after the end of the last glaciation, as
Antarctica slid toward the pole, but not before it had time to
pass on its wisdom and knowledge of mathematics and the
sciences to the peoples who'd in due course become the
Aztecs, Olmecs, Maya and Egyptians:

> . . . our species could have been afflicted with some terrible
> amnesia and . . . the dark period so blithely and dismissively
> referred to as "prehistory" might turn out to conceal some
> unimagined truths about our own past.

It might indeed. Prehistory most certainly represents a large
gap in our knowledge of events, which is why it got its name.
This is no reason, however, to believe the first person to
come along offering a narrative that can be used to fill in
the hole; to do so would be to fall for the god-of-the-gaps
fallacy (see page 20).

The destruction of civilization marked by the disappearance of the Atlanteans – which Hancock dates using
means such as "precessional dating", his own invention, to

10,500BC – is a part of a cycle of such destructions, of which the next is due on December 23 2012, a date that must seem uncomfortably less distant to Hancock now than it did in 1995. The knowledge of this doomsday was coded by the Atlanteans into monuments like the Great Pyramids; unfortunately, it's only today that these monuments are beginning to be decoded by people like Graham Hancock, giving humanity far too short a warning to enable us to avert the apocalypse.

In a major later book, *Underworld* (2002), Hancock appears to have abandoned, or at least set temporarily to one side, his interest in Antarctica as Atlantis, instead looking – with the aid of his newfound hobby of scuba diving – for his ancient civilization in Malta, in India, and in the region of Japan, China and Taiwan. His thesis here is that, using a system of computer modelling developed by Glenn Milne of the University of Durham, UK, we can see how coastlines looked at various stages in the past; we can thus find dates for the inundation of coastal settlements. The flaw in his method is that Milne's models are of limited and quite specific use: they allow only for coastline changes brought about by variations in sea level; they don't take account of instances where, perhaps through tectonic activity, land slips into the water. For example, a past event of the order of the anticipated "Big One" along the San Andreas Fault plunging large parts of northern California into the Pacific would not register on Milne's maps unless it were known about from some other source.

That communities do become inundated is not a matter of controversy – all over the world there are examples of coastal settlements that have been lost to the sea through erosion and/or rising sea levels. A fine example of this occurred around the Baltic Sea between about 6100BC and 3400BC in the wake of the last glaciation and the resultant breaching, between Germany and Denmark, of what had once been a large freshwater lake by the waters of the Atlantic. A German project called SINCOS (more fully, The Sinking Coasts: Geosphere, Ecosphere and Anthroposphere of the Holocene Southern Baltic Sea) has been diving to retrieve artefacts and other remains from the communities submerged as the water level rose by about 8m; even though the total rise took millennia, it appears the first 3.5m or so happened in a period of mere days, so that much is preserved underwater in the fairly natural state it was in when the inhabitants had hastily to abandon it. Of course, what the SINCOS scientists are revealing is not the relic of a prehistoric technological superstate, rather a set of Neolithic communities that lived largely through fishing.

In Malta, after what sounds like a frustrating encounter with Hubert Zeitlmair (see page 127), Hancock found the Mifsuds and their colleagues much more rewarding, accepting their unorthodox hypotheses about the Maltese Atlantis more or less wholesale. In India he was much influenced by that school of mythological interpretation which maintains that the vedas are far closer to historical truth than orthodox archaeology allows. He posits a coastal civilization of even greater antiquity than those known to have flourished in India, then sets out to find remnants of it at submerged sites like the one just offshore from Mahabalipuram, in southeast India's Tamil Nadu state. Although he accepts the Mahabalipuram ruins are dated with fair confidence to the 7th century, Hancock plays with the notion that they might be far older – as old as 3000BC, perhaps. This is still very, very much younger than the kind of date required for a primeval lost civilization. A similar cavalier attitude towards dating is evident in Hancock's treatment of what appear to be two submerged cities discovered in the Gulf of Khambhat (Cambay), western India, in 2000; using Milne's models he dates the site to 5700BC, even though there's no reason to believe the underwater cities, *if they even exist*, were not victims of a much more recent land slippage.

In this instance he has some support from some of the members of the Indian National Institute of Ocean Technology (NIOT), the authority investigating the site, and even from some senior Indian politicians eager to promote tourism. The region involved is much affected by currents, and the turbulence stirs up such quantities of mud that direct observation is impossible. NIOT has therefore been carrying out its investigations using techniques like sonar – to detect the patterns of apparent structures under the sludge of the sea floor – and dredging, to trawl for anything that might look like an associated artefact. Many

Indian geologists assert that the geometrical patterns observed by sonar and assumed by NIOT to be city streets and buildings are in fact natural formations, the impression of artificiality being compounded by artefacts of the sonar imaging process itself. The status of the recovered objects is equally controversial: while some involved have claimed that, for example, pottery shards date back as far as 13,000 years ago – which really *would* be sufficiently ancient for Hancock's purposes! – others have pointed out that this is exactly the same date given by testing the sediments in which the supposed pottery shards were found, so that the "pottery" is more likely a product of sedimentary compaction.

Farther east, Hancock travelled to Yonaguni, which is the westernmost island of Japan, and to Akajima, one of the Kerama Isles in Japan's Okinawa Prefecture. Although some – notably marine scientist Masaaki Kimura (b1940) of the University of the Ryukyus in Okinawa – maintain that the so-called Yonaguni Monument, offshore from the island, is the remnant of an ancient city, most reckon it to be a natural formation. Off the coast of Akajima, near Tom Moya Reef, there was discovered in 1995 an unusual submarine stone circle that very much resembles ancient megaliths like Stonehenge; while this may again be a natural formation it's possible that, since Japan's Jomon culture did erect stone circles, we indeed have here an example of a piece of very old architecture . . . but it's hardly the product of an advanced technological civilization.

The book *Underworld* was associated with a Channel 4 television series of the same name, and it is reported that it was at Channel 4's insistence that Hancock made a better case than in his previous works, rather than merely rehashing his old conclusions, and that he also allowed space for contrary views. (This is one of the reasons why *Underworld* is such an enormous book, weighing in at nearly 800 pages.) It seems even Hancock recognized this was an improvement, because he posted on his website that

> [when writing *Fingerprints of the Gods*] my top priority was to cram in and get down on the page anything and everything that I thought might weigh in favour of the lost civilization

idea. This was more important to me at that time than taking meticulous care with the quality of every source or being choosy about what leads I followed. I was also too quick to attack weaknesses in the orthodox position while failing to take proper account of orthodox strengths. The result was that my case for a lost civilization was anything but bullet-proof, and *Fingerprints* has come in for a massive amount of criticism – some of it richly deserved.

In November 1999 BBC Television's flagship science programme *Horizon* broadcast an episode called *Atlantis Reborn* that examined in depth Graham Hancock's unorthodox archaeology. In it he pointed out:

> I avoid using the word "Atlantis" in my books because most people, when they hear the word "Atlantis", immediately think that they're dealing with the lunatic fringe. I don't feel that I belong to a lunatic fringe.

After much analysis of Hancock's theories concerning the Pyramids, the Sphinx and the Bolivian site Tiwanaku, the programme turned in its closing stages to his belief that Yonaguni is a 12,000-year-old city. In order to bolster his hypothesis, Hancock invited the geologist Robert M. Schoch, of the University of Boston, to join him in diving at the site. Schoch, no stranger to archaeological controversy (see page 158-9), was disappointed:

> I went there in this case actually hoping that it was a totally manmade structure that was now submerged underwater, that dated maybe back to 6000BC or more. When I got there and I got to dive on the structure I have to admit I was very, very disappointed because I was basically convinced after a few dives that this was primarily, possibly totally, a natural structure. . . . Isolated portions of it look like they're manmade, but when you look at it in context – you look at the shore features, etc. – you see how, in this case, fine sandstones split along horizontal bedding planes that gives you these regular features. I'm convinced it's a natural structure.

The *Horizon* episode, while debunking Hancock's hypotheses, did not give him an especially rough ride. Nonetheless,

Hancock and his ally and parallel researcher Robert Bauval (b1948) complained bitterly to the Broadcasting Standards Commission that they and their theories had been unfairly treated. And, if you read Hancock's website, you will conclude that the BSC upheld their complaint.

The truth is not exactly thus. In fact the BSC did conclude that *Horizon* had given unfair coverage to the Bauval/Hancock hypothesis that the three Great Pyramids were laid out in 2500BC in imitation of the way the three stars of Orion's belt would have looked in 10,500BC (see page 150), in that the producers had omitted to include the two men's counterarguments to the hypothesis's rebuttal by astronomer Ed Krupp. On the other ten points of complaint, however, the BSC found in favour of *Horizon* – a matter Hancock deems undeserving of mention.

And some of those other nine concerns were more substantial than an academic disagreement over the reason for the layout of the Great Pyramids. Here's one, drawn from the BSC's synopsis of its adjudication:

> The programme had created the impression that he was an intellectual fraudster who had put forward half-baked theories and ideas in bad faith, and that he was incompetent to defend his own arguments.
>
> *Adjudication:* [The Commission] finds no unfairness to Mr Hancock in these matters.

By contrast with Atlantis, the legend of Lemuria – the great lost continent of the Pacific – was born in scientific speculation. The geologist Philip Lutley Sclater (1829–1913), puzzled by the modern distribution of lemurs, suggested in 1855 there could have been a now-lost landmass in the Indian Ocean over which the ancestors of today's lemurs migrated. What had persuaded Sclater to think in terms of lost lands was the (rumoured) behaviour of lemmings: obviously the reason they swim to their deaths from Europe's western coasts is that they're trying to migrate to Atlantis.

A couple of decades later, likewise musing the lemur problem, the German naturalist Ernst Heinrich Haeckel (1834–1919) came to the same conclusion, perhaps inde-

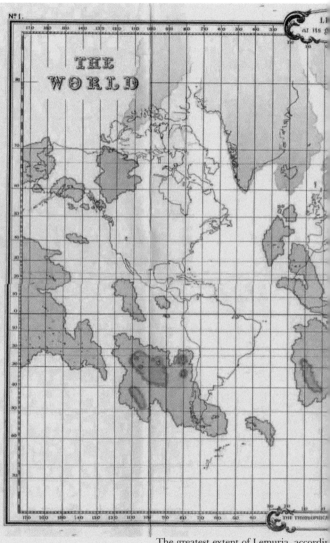

The greatest extent of Lemuria, accordi

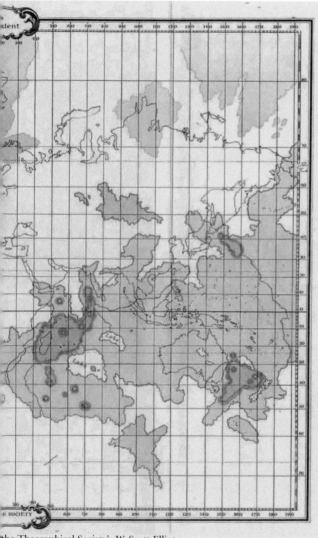

the Theosophical Society's W. Scott-Elliot

pendently. Unfortunately Haeckel couldn't leave it there, and went on to add that the Aryan race must have originated in Lemuria before settling in Asia. So fixated was he on this profoundly nonscientific notion that he published a map of the Lemurians' migratory routes in his book *The History of Creation* (1879).

Once Helena Blavatsky (1831–1891) got her hands on Lemuria, the floodgates were thrown open to all sorts of lunacy. The Lemurians, she revealed without the slightest trace of corroborative evidence, were at first giant apelike hermaphrodites but slowly evolved away from this idyllic state to become more like modern humans. The destruction of Lemuria was somehow linked to its inhabitants' discovery of sex; luckily a few of them escaped to Atlantis before catastrophe befell the Lemurian continent, and there they dwelled in relative safety . . . little realizing they'd gone from frying pan to fire. Rudolf Steiner (1861–1925) added, again *sans* evidence, that the Lemurians did not possess the power of reason but were able to get along fine without it, being blessed with highly efficient instinct. In addition, they were telepathic and psychokinetic.

An important early item in the US literature on the subject was Wishar S. Cervé's *Lemuria: The Lost Continent of the Pacific* (1931). This book, published by the Rosicrucians (AMORC) of San Jose, California, was in fact written by Harvey Spencer Lewis (1883–1939), founder and first Imperator of AMORC. According to Lewis, while the rest of the Native Americans were descendants of the Ten Lost Tribes of Israel – a not uncommon hypothesis of the time – the Maya were descended from survivors of the lost continents of Atlantis and Lemuria. The Maya weren't the only ones! If you were lucky enough you might see evidence of the Lemurian survivors who still today dwelt on the slopes of Mount Shasta, northern California: mysterious lights at night, odd boat-shaped airships and all. And if you were *really* lucky you might meet an actual Lemurian on Shasta; you'd be able to tell him by the long hair, the English accent and the angelic mien.

It was probably Lewis's book that sparked off one of the more famous newspaper hoaxes related to bogus science. On Sunday May 22 1932 the *Los Angeles Times* published an article, "A People of Mystery", by reporter Edward Lanser. Lanser told of being aboard a train that passed close by Mount Shasta and seeing a mighty light upon the slopes. He asked the conductor what this could be:

> "Lemurians," he said. "They hold ceremonials up there."
> Lemurians!
> The fact that a group of people conduct ceremonials on the side of a mountain is not of exceptional interest, but when these people are said to be Lemurians, that is startling . . .

Too true. Eagerly Lanser investigated among the inhabitants of the region and discovered that the existence on Shasta of a "mystic village" of Lemurians was common knowledge. Occasionally the Lemurians, who looked just like Lewis said they did, would visit the local towns in order to buy supplies with gold nuggets. But they'd been subjected to scientific study:

> . . . I learned that the existence of Lemurian descendants on Mt. Shasta was vouched for some years ago by no less an authority than the eminent scientist, Prof. Edgar Lucien Larkin, for many years director of the Mt. Lowe Observatory in Southern California.
> Prof. Larkin, with determined sagacity, penetrated the Shasta wilderness as far as he could – or dared – and then, cleverly, continued his investigations from a promontory with a powerful long-distance telescope.
> What the scientist saw, he reported, was a great temple in the heart of the mystic village – a marvelous work of carved marble and onyx, rivaling in beauty and architectural splendor the magnificence of the temples of Yucatán. He saw a village housing from 600 to 1000 people; they appeared to be industriously engaged in the manufacture of articles necessary to their consumption, they were engaged in farming the sunny slopes and glens surrounding the village – with miraculous results, judging from the astounding vegetation revealed to Prof. Larkin's spy-glass. He found them to be a peaceful community, evidently contented to live as their ancient forebears lived before Lemuria was swallowed up by the sea.
> When Prof. Larkin concluded his investigation he had gathered enough proof to warrant him to say that in this village, in a secluded glen at the foot of Mt. Shasta's partially

extinct volcano, far from the beaten paths that lead to our civilization, there live the last descendants of the first inhabitants
of this earth, the Lemurians.

Surprisingly, although Edgar Lucien Larkin (1847–1925)
seems to have had no interest in Mount Shasta, he did in
fact exist, and was indeed the "director" of the Mount Lowe
Observatory. The observatory, however, was not quite on a
par with its more famous cousins on Mounts Wilson and
Palomar. Instead, it was a tourist attraction attached to the
Echo Mountain House hotel, served by the Mount Lowe
Railway; although some serious astronomy was done using
the observatory's 16in (410mm) telescope, Larkin's main
task was to show guests the moon and stars through it. He
received greater notice as a mystic, publishing three books
including *Matchless Altar of the Soul* (1916).

In her *Secrets & Mysteries of the World* Sylvia Browne
clearly regards Lemuria and Mu as merely different names
for the same continent. And she offers us some revealing
glimpses, psychically derived in her usual incontrovertible
fashion, of life among the Lemurians. They all dressed very
much alike, and their houses were all very much alike, too:
pyramidal. The Lemurians were interested in growing fruit
and vegetables "the size of most people's heads". They liked
"herbs, naturalistic and holistic cures, and the laying on of
hands". All in all, life on Lemuria sounds pretty dismal and
dreary – except for devotees of vegetable marrow contests.
There must have been many inhabitants of the doomed
continent who breathed a heartfelt sigh of relief when the
ground began to tilt . . .

We've inherited, Browne maintains, a few cultural items
from the Lemurians, notably the Sanskrit language. Indeed,
we'd know far more about the lost continent if only more of
us could – and were allowed to – read secret Sanskrit documents that detail many aspects of this prehistoric civilization. Browne adds, although she doesn't support the claim:
"The Essenes and the entire Gnostic movement also date
back to this time."

In this book it's surprising to find Browne still, in 2005,
holding out hope for the "architecture" and "roadways"
observed during a 1968 aerial survey of the sea floor

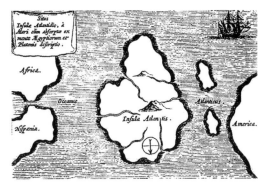

Map of Atlantis done in the 17th century by Athanasius Kircher

offshore from Bimini, Florida, even though follow-up investigations using divers soon enough showed that the "architectural" objects were made of 19th-century concrete (how
they came about has still not been definitively explained)
and the "roads" were natural limestone formations. This is
a characteristic of archaeological information transmitted
from the spirit world: it seems very often to be a bit out of
date, almost as if derived from elderly magazine articles.
Perhaps time functions differently in the afterlife.

Browne has not been alone in being able to bypass
archaeology in her quest for lost civilizations. According to
Edgar Cayce (1877–1945), the end of the world will come
about not through climate disaster or nuclear Armageddon
but through the rise of Atlantis from beneath the waters of
the Atlantic Ocean, where it sank some 12,000 years ago.
The Atlanteans were, after all, the first to possess nuclear
bombs, death rays and robots/computers. It was here that
humankind first chose to assume corporeal form, having
before this been purely spiritual. Because our species has
become so vicious, the Powers That Be have decided we
must be exterminated, and the selected agents of that exter-
mination are the Atlanteans, emergent from the ocean and
armed to the teeth. This is all going to happen by the year
2000.

Correction: This *was* all going to happen by the year 2000.

Throughout his life Cayce added to our information about Atlantis, releasing "information" in dribs and drabs, often in the context of discussing previous incarnations there. The continent stretched all the way from the Azores to the Sargasso Sea. Civilization flourished there from about 200,000BC until the last of a series of catastrophes doomed Atlantis a little after 10,000BC. The Atlanteans derived energy from gas, electricity and steam, but most importantly from a kind of crystal that could be used to trap solar energy. And so on, interminably on.

Perhaps the Atlanteans weren't human at all, but came originally from the stars.* In *One Foot in Atlantis* (1998) William Henry tells us about Nicholas Roerich, an agent of the Serpent Brotherhood who was revered by Franklin Delano Roosevelt (1882–1945) and Henry A. Wallace (1888–1965), whose support he garnered to search for Christ's body in Shambhala. It seems Wallace and Roosevelt were persuaded that various spiritual masters had survived the destruction of Atlantis to make a new and secret home in Shambhala, whence they covertly influenced world affairs; naturally, the two patriots wanted to strike up an alliance between the US and Shambhala. Roerich had proof of his claims in the form of a fragment of a mysterious extraterrestrial stone that had once belonged to Solomon and before that to the Emperor of Atlantis, who'd got it from people from Sirius; it's no coincidence that Sirius is in the approximate direction of the galactic core, which is more properly known as Thule or Hyperborea – as the Greeks knew. Later interpreters have erroneously thought the Greeks believed Thule/Hyperborea to be in the far north, but the Greeks were *not that silly*. The rest of the mysterious Sirian stone was even now in a tower in

---

* In *Mystery of the Ancients* (1974) Craig and Eric Umland claimed exactly this of the Maya, who came from another solar system and initially made their base on the fifth planet, between Mars and Jupiter. When this exploded (forming the asteroid belt), the Maya reluctantly found a new home on earth. Once here they displayed all the signs of advanced civilization, such as ritual human sacrifice to the gods. Of course, since some atlantologists think the Maya were Atlantean survivors . . .

Shambhala radiating rays yet more mysterious than itself which shaped the course of civilization.

Browne informs us, again in her *Secrets & Mysteries of the World*, that there's some dispute as to the origins of the extraterrestrials who came here and founded the Atlantean civilization: some people think they came from "the Lyrian star system" – a planet of a star in the constellation Lyris, perhaps? – whereas Francine, Browne's spirit medium, "says that it was Andromeda". From a planet of a star in the constellation of Andromeda? Or from a much more distant place, the Andromeda galaxy? The text offers no clarification except what could be construed as a tantalizing hint: "This possibly explains why Atlanteans could levitate."

Or maybe people we just *think* are visitors from other stars are in fact surviving Atlanteans? According to the leading flat-earther Samuel Shenton (see page 50), at some time in the remote past the world was torn apart. By the time things settled, one of its continents had been subsumed beneath the north pole (which in Shenton's cosmology lay at the centre of the terrestrial disc). That continent, Atlantis, was not destroyed, however; indeed, we can still witness the flying saucers its inhabitants send out to survey the rest of the world.

Some people really believe the Giza Pyramids hold the key to the future.

In 1859 the UK publisher John Taylor (1781–1864) published his book *The Great Pyramid: Why Was it Built & Who Built It?*, in which he deduced that the Egyptians had constructed the Pyramid as a representation of our planet – building a sphere of appropriately enormous size obviously being beyond feasibility.

Taylor also calculated, following on the heels of Agatharchides of Cnidus (2nd century BC), that the base length of the Great Pyramid represented exactly one-eighth of an arc-minute of the earth's circumference. The base length of the Great Pyramid is 230.33m. Performing the necessary multiplications, we deduce a figure for the earth's circumference of about 39,800km. The equatorial circumference of the earth is 40,076km, its polar circumference 40,008km, so here at least the match was impressively close . . . except that the argument that the Great Pyramid's builders might have deliberately created the monument to have a base length of one-eighth of an arc-minute of the earth's circumference is surely weak; why would they have chosen this particular arbitrary ratio?

Whatever the truth of the mathematics, Taylor's deductions were largely ignored by his contemporaries because of his insistence that the only way the Egyptians could have known about such things as the earth's circumference was if God had spoken directly to them. Although the impact of *The Origin of Species* (1859) had still to make itself fully felt, even the largely religious scientists of the time found the notion of a Divine Revelation to the Egyptians too much to swallow. Taylor further concluded that the architect of the Great Pyramid had been the Biblical Noah.

If Taylor was sufficiently staunch in his faith as to contemplate the direct intervention of God as a plausible historical event, he was matched by his friend Charles Piazzi Smyth (1819–1900), the Astronomer Royal of Scotland. Smyth made his own examination of the Great Pyramid in the early 1860s, publishing his conclusions in *Our Inheritance in the Great Pyramid* (1864; later expanded editions were titled *The Great Pyramid: Its Secrets and Mysteries Revealed*). From his studies Smyth concluded that the ancient Egyptian architects had adhered to a unit of measurement which he called the Pyramid Inch. About 1.001 times the size of our modern inch, this measure had, according to Smyth, been used by the architects in various extraordinarily significant ways; for example, he counted the number of Pyramid Inches in the Great Pyramid's perimeter and claimed it was exactly 1000 times the number of days in the year, and there was a further numerical relationship between the monument's height, in Pyramid Inches, and the distance from the earth to the sun, in miles – a relationship all the more curious because the ancient Egyptians knew nothing of miles. Exactly ten

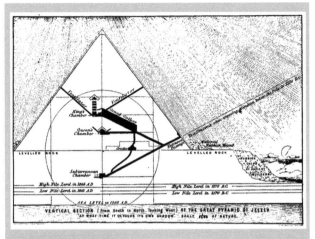

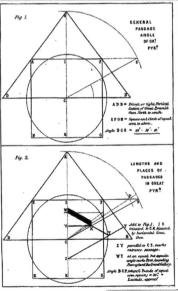

Drawings done by
Charles Piazzi Smyth in
support of his many
numerological
speculations concerning
the Great Pyramid

million Pyramid Inches, Smyth calculated, gave you the earth's polar radius. And so on. One of Piazzi Smyth's more astonishing numerical discoveries was that, if you calculate the Great Pyramid's volume in cubic inches, the result is equal to the number of human beings who have ever lived. Since he took as his starting point the Creation, just a few thousand years ago . . .

Even more excitingly, following in the footsteps of a devout crank called Robert Menzies,* Smyth was able, using the Pyramid Inch, to discover that the Great Pyramid was a glorious library of prophecies and revelations concerning the earth's history; measurements revealed to him that the world had indeed been created in 4004BC, that the Flood could be dated to 2400BC, and that the Second Coming would be in 1911, or perhaps 1920, or maybe even 1925 . . . In fact, the great embarrassment was that the first of the dates Smyth calculated from Great Pyramid measurements as the end of the world was 1874, which year passed uneventfully while he was still alive.

Another blow concerned the Pyramid Inch, which measure Smyth had derived from the supposedly standard size of the Pyramid's casing stones. Later investigators found these stones to be of all different sizes.

There was a further, political implication of the Pyramid Inch. Smyth had sympathies with the ideas put forward by the eccentric John Wilson (1799–1870) in *Our Israelitish Origin* (1840), the book indirectly responsible for the British Israelite movement. The almost exact equality between the Pyramid Inch and the familiar inch demonstrated, if any further demonstration were needed, that the British were descended from a people who'd been present in ancient Egypt – i.e., the Israelites. The inch, and by extension the Imperial system of units, could thus be deduced as God-ordained. This was one of the rationales Smyth offered in his scholarly papers for urging European scientists to give up their godless metric system.*

* Amusingly, a number of later pyramidologists assume this Robert Menzies, a Scot of whom not much is known except that he announced his pyramidological findings about 1865, was the Australian prime minister Sir Robert Gordon Menzies (1894–1978).

But there was more. According to Smyth, the Great Pyramid represented a squaring of the circle: the base area of the edifice was the same as that of a circle whose radius was the pyramid's height. Earlier, John Taylor had said the dimensions of the Great Pyramid represented the Golden Section. Both claims relied upon the angle of slope of the pyramid's side, something that couldn't be determined with exact precision because the construction had lost its outer casing over the millennia. Even so, it could be very well estimated at 52°, and Piazzi Smyth's argument seemed convincing: the slope to render the squaring-circle result was 51° 51.2′, which was as near to 52° as one could hope for.

But then it was realized that the slope for Taylor's "Golden Section" result would be 51° 49.6′, which was *also* extremely close to 52°!

Only later did people start to wonder how these angles might have appeared to an ancient Egyptian. The Egyptians used a different system of angular measurement from the one we're accustomed to. Without going into detail, in the view of an ancient Egyptian, if the slope of the Pyramid were a tidy angle of 5½ palms the building's height would be an equally tidy 280 ells. And an angle of 5½ palms equals what we'd call 51° 50.6′ – which is *again* extremely close to the estimated 52°.

What really put the dampers on both Piazzi Smyth's and Taylor's theories was the discovery of the *Papyrus Rhind*, which is essentially an ancient Egyptian school maths textbook. In this it was found that the ancient Egyptians thought pi was 3.1604 – not a huge difference from π's true

* This opposition to the metre was one of the reasons pyramidology found such fertile ground in the US. As example, the Ohio Auxiliary Society (which was supported by President Garfield) published a journal called *The International Standard* to defend against the metric system: "We believe our works to be of God; we are actuated by no selfish or mercenary motive. We deprecate personal antagonisms of every kind, but we proclaim a ceaseless antagonism to that great evil, The French Metric System. . . . May our banner be ever upheld in the cause of Truth, Freedom, and Universal Brotherhood, founded upon a just weight and a just measure, which alone are acceptable to the Lord."

value of 3.14159..., but enough to make a nonsense of speculations about the Pyramid representing a squaring of the circle.

In June 1876 Smyth published an article on his theories in the US journal *Bible Examiner*, and a few years later the publisher of that journal, George Storrs (1796–1879), wrote a series of articles about the Great Pyramid and its significance in prophecy in the Adventist journal *Herald of Life and the Coming Kingdom*. The relationship between pyramidology and cultist Christianity in the US was thereby cemented. In this context, the most important convert to pyramidological ideas was Charles Taze Russell (1852–1916), first President of the Watch Tower Bible and Tract Society and founder of the organization that would eventually become the Jehovah's Witnesses. In 1883 he announced that God Himself had created the Great Pyramid as a sign. And in his book *Thy Kingdom Come* (1897) he devoted an entire chapter to the subject of the Great Pyramid. Adopting a chronology from the Adventist Nelson H. Barbour (1824–1905), Russell regarded the edifice as being the "divine plan of the ages in stone".*

It was numerology that attracted men like Russell and Barbour to pyramidology. After years of attempting to establish numerologically the date of the Second Coming through analysis of sacred texts like *Revelation*, it was only natural that they should accept eagerly a scheme like that of Smyth, who found significant numbers in the pyramid everywhere he looked. Moreover, he was a *scientist*, and Russell and the other leaders of the Adventist movement regarded themselves as being likewise – hard though it might be for us to imagine this integral mixture of belief in the sky's imminent collapse and reverence for natural science. In *Thy Kingdom Come* Russell exemplified the attitude:

> The Great Pyramid, however, proves to be a storehouse of important truth – scientific, historic, and prophetic – and its testimony is found to be in perfect accord with the Bible, expressing the prominent features of its truths in beautiful and fitting symbols. It is by no means an addition to the written

* In 1928 Russell's successor, Joseph F. Rutherford (1869–1942), rejected pyramidology as the work of the Devil.

revelation: that revelation is complete and perfect and needs no addition. But it is a strong corroborative witness to God's plan; and few students can carefully examine it, marking the harmony of its testimony with that of the written Word, without feeling impressed that its construction was planned and directed by the same divine wisdom, and that it is the pillar of witness referred to by [Isaiah].

There seems to be little logical difference between Russell's religion-based faith in the divine construction of – or at least inspiration for – the Great Pyramid and the ideas of more recent thinkers that the monument was built by survivors from Atlantis as a repository for the wisdom of their culture or by extraterrestrial visitors for similar purposes. The only real distinction is in the shifting description of the deity.

Some people really believe the Giza Pyramids were astronomical observatories.

In *The Dawn of Astronomy* (1894) the astronomer Norman Lockyer (1836–1920) made a forceful argument that the Pyramids and other ancient Egyptian structures were oriented in such a way as to facilitate the making of astronomical observations, presumably related to calendrical calculations such as the determination of the solstices. Curiously, Lockyer's was far from a new notion: centuries earlier various Greek and Arabian writers had offered the same speculation, and certainly the ancient Egyptians were exceptionally interested in the stars.* The constellation we call Orion was to them known as Sahu and was supposedly the home of the dead. It seems Orion and its neighbour, the bright star Sirius, held especial significance to the Egyptians because their annual rising served as a kind of clock coordinating with the flooding of the Nile.

Plenty of more recent workers have developed

---

* Although, and all due credit to them, it seems they never invented astrology.

Lockyer's ideas. Various shafts within the Pyramids, initially thought to have been merely ventilation shafts created to allow the labourers to breathe and blocked off as construction approached completion, have been reinterpreted as being also portals through which specific portions of the skies could be observed. Establishing which portions those might have been and then calculating their angular position in the sky in past eras, taking into account the effect of equinoctial precession, gives a way of dating the Pyramids' construction. For example, one of the shafts within the Great Pyramid would have pointed at the pole star during the period 3000–2400BC, which happily embraces the accepted date for when the monument was built – during the Fourth Dynasty, which lasted 2613–2494BC. It's perfectly possible, of course, that such correlations are merely coincidence and that the shafts were indeed incorporated solely for ventilation.

Is it feasible that the Giza Pyramids were laid out to reflect a pattern the builders saw in the sky? This is the contention of authors Robert Bauval and Adrian Gilbert in their book *The Orion Mystery* (1994); the notion has since received the enthusiastic backing of Graham Hancock. Bauval and the others observe that the three Giza structures form a line diagonal to north–south, with the smallest of the three being displaced slightly away from what would otherwise be a straight line. This, they point out, is startlingly reminiscent of the arrangement of the three stars in Orion's belt, which form an approximate straight line. It hardly needs to be stated that such a coincidence of pattern with a mere three points is hardly spectacular: by definition, two points form a straight line and any third point placed randomly between them is likely to be offset to some extent or other from that straight.* Mindful of this, Bauval and

---

*An argument used to rubbish Bauval's contention by Ed Krupp and others is that the supposed depiction of Orion's belt formed by the Pyramids is upside-down from the way the belt is seen in the sky – i.e., the offset of the central point is in the wrong direction. This strikes me as specious. There's not really a right way up for laying out a stellar pattern on the ground.

Gilbert also bring into the picture the non-Giza pyramids of the pharaohs Nebka and Djedefra, which they say are extensions of the design to represent two of Orion's other stars. Here there's still an approximate match, but it's only approximate.

There are more mundane explanations for the layout of the three Giza Pyramids, involving two practical matters: the terrain, and the suitability of the site chosen for each as a place from which the architects could get a clear view of north in order to align the pyramid's sides accurately north–south and east–west. When the second of the three edifices, that of Khafre, was planned, its position had to take account of the fact that the quarrying to build Khufu's pyramid had left a huge hole in the ground. The third pyramid to be built, that of Menkaure, couldn't be built in a straight line with the other two because the land sloped steeply where it would have had to be built. The builders settled for the best approximation they could sensibly get.

This is not to say that the Bauval/Gilbert suggestion is impossible; simply that it seems a lot less likely than the argument from practicability essayed by orthodox egyptology.

Some people really believe archaeologists have it all wrong about the age of the Pyramids. While most people who suggest a date for the Pyramids other than the accepted one are trying to argue the structures are far older, this is not universally the case. Immanuel Velikovsky produced a revised and much truncated chronology in his book *Ages in Chaos* (1952) and its successors. Much more recently, in *The Empire of Thebes, or Ages in Chaos Revisited* (2007) and its companions in the *Ages in Alignment* series, the Irish author Emmet Sweeney, a lecturer in modern history at West University in Timisoara, Romania, likewise argues they were built much later than generally thought:

. . . the accepted chronology of ancient civilizations has hitherto been out of alignment, out of joint. Kingdoms, empires and individuals in the different regions and cultures of the ancient Near East have been set down in the history books in a chaotic order, with the result that kings in Israel, for example, who were contemporary with pharaohs of the Egyptian New Kingdom (18th and 19th Dynasties) have been placed centuries after those same pharaohs. Thus the "history" of the ancient East, found in the voluminous learned textbooks of the great libraries of the world, is a fiction. . . .

Bringing the 18th Dynasty down into the 7th century [BC] has of course dramatic implications for the whole of ancient history. For example, Akhnaton cannot be assigned to around 610BC if the pyramid builders are left in the 3rd millennium. The evidence suggests that this epoch too, whose artisans cut iron-hard granite with mathematical precision and employed advanced Pythagorean-style geometry in their designs, must be reduced by a margin even greater than that of the 18th Dynasty . . .

Other neo-Velikovskian revisions of ancient chronology include that by Peter James set out in *Centuries of Darkness: A Challenge to the Conventional Chronology of Old World Archaeology* (1991), which is archaeologically based and has attracted a fair amount of learned debate, and John J. Bimson's in *Redating the Exodus and Conquest* (1978), which is based on Biblical analysis.

The dating of the Great Sphinx is for obvious reasons very much tied in with the dating of the Pyramids. It's generally accepted the Sphinx was almost certainly built at about the same time as its neighbours, probably during the reign of Khafre (or Khafra; reigned perhaps 2558–2532BC), but repairs and restorations carried out at various periods of history complicate the picture.

In the late 1980s and early 1990s researches done by the University of Boston geologist Robert M. Schoch led him to propose that, while the granite facing of the Sphinx might well have been added c2500BC – i.e., during the pyramid-building era – the limestone core of the sculpture may have been constructed much earlier, perhaps as early as 7000BC although more likely around 5000BC. He based this opinion on the fact that the core appears to have been very heavily eroded, which erosion he attributed to rains;

it's because rain hasn't been a prominent feature of Egyptian life for several thousand years, but the area *was* subject to heavy rains during the period 5000–3000BC, that Schoch suggests the much earlier date. Conventional archaeologists maintain the erosion was more likely caused by windblown sand, and could have occurred at a sufficient rate to match the accepted construction date. Schoch offers his views on the ancient world in such books as *Voices of the Rocks: A Scientist Looks at Catastrophes and Ancient Civilizations* (1999) and *Pyramid Quest: Secrets of the Great Pyramid and the Dawn of Civilization* (2005).

It's easy to confuse Schoch's hypothesis about the age of the Sphinx with that of John Anthony West, whose website boasts of his "symbolist Egyptian research" and offers "Magical Egypt" tours. The author of books like *Serpent in the Sky: The High Wisdom of Ancient Egypt* (1979) and *The Case for Astrology* (1970; with Jan Gerhard Toonder), West believes that the supposed rain and flood erosion of the Sphinx's base indicates a *much* earlier date of construction, 9000BC. This would link in with the noachian Flood, which West equates with the waters released at the end of the last glaciation. It's his contention that the builders of the Sphinx were refugees from Atlantis.

Graham Hancock (see pages 136ff) likewise believes that refugee Atlanteans built the Sphinx, and also the Pyramids, the period of construction being even earlier than West posits, about the middle of the 11th millennium BC. The Sphinx's body took the form of a lion because constructed during the Age of Leo, which lasted approximately 10,970–8810BC. Hancock traces other symbolisms to the different Ages defined by the precession of the equinoxes; for example, the Age of Pisces, recently ended, was marked by the ascendance of Christianity, one of whose symbols is the fish, while the Age of Aries saw the popularity in Egypt of the ram god Amon, not to mention the ram sacrifices mentioned in the Old Testament. (And perhaps, to continue his line of thought, the Age of Aquarius will be marked by widespread flooding as the glaciers melt.)

One objection to the linking of the Sphinx's form with its construction during the Age of Leo is that all the available evidence points to the fact that the zodiacal constella-

tions, Leo included, weren't given their present identities
until sometime around the second millennium BC – thousands of years after Hancock supposes the Sphinx was built.

In *Voyages of the Pyramid Builders: The True Origins of the
Pyramids from Lost Egypt to Ancient America* (2003) Robert M.
Schoch, whom we've just met, contends that a civilization of
pyramid builders lived in the Sundaland, the land area
posited once to have existed between Indonesia and
Southeast Asia, but were driven from there in perhaps
6000BC to other parts of the world when global rises in sea
level (caused, in Schoch's scenario, by cometary activity)
submerged their territory.

According to Edgar Cayce the annals of Atlantis – the
so-called Akashic Records – are concealed in a subterranean
Hall of Records that lies somewhere between the Great

Edgar Cayce

Pyramid and the Sphinx and will
be discovered during the 1990s.
Will be? Oops.

Leaving aside the date of
discovery, the matter is of some
importance to Caycean archaeology. In dribs and drabs, Cayce
told a story of Egypt's earliest
history, through the figure of the
priest Ra-Ta, that is in discord
with the findings of conventional
archaeology. Ra-Ta joined with a
group from the Caucasus who
went to Egypt and found there a
ragbag of fairly unsophisticated
cultures. They allied with the
most advanced of these and
began to establish a fairly benign civilization. Soon this society was joined by a refugee people from Atlantis. After
certain political upheavals and a civil war, Ra-Ta was
appointed absolute ruler and set about doing things like
build the Great Pyramid, with Hermes as architect and
Hept-Sept as overseer. All of this happened in about
10,500BC.

A whole array of scientific techniques show quite clearly
that civilization sprang up in the region around 4500BC,

over the centuries gradually progressing from hunter-gathering to the stage where people could build pyramids. Further, there's no mention anywhere in ancient inscriptions of Ra-Ta or any of the other characters in Cayce's account, which surely there should have been bearing in mind the importance to Egyptian civilization of these individuals. There is, however, hard and fast evidence within the Great Pyramid that it was constructed in the latter part of Khufu's reign.

All might have seemed lost for Caycean archaeology to anyone save Hugh Lynn Cayce (1907–1982), Edgar's eldest son, who persevered in claiming that his father had the right of it and archaeology was wrong. However, an anthropology student working (for fear of academic retribution) under the pseudonym Richard Roche reanalyzed Cayce's material and was struck by the resemblance between the story of Ra-Ta and that given by Plutarch as the story of Osiris. In *Egyptian Myths and the Ra-Ta Story* (1975) Roche argued that Egypt's First Dynasty, dated to 3100–2890, was not really a first dynasty at all but represented a *reunification* of a much older civilization that had disintegrated in earlier centuries; what most people took to be Egyptian mythology was in fact, in Roche's view, a garbled tale of a genuine history.* If only the Hall of Records predicted by Cayce could be excavated, Roche stated, proof of the kind of chronology Cayce's account demanded would surely be forthcoming. Failing that, since conducting major excavations under the Sphinx is something difficult to do, Roche urged that excavation be done at the Abu Simbel-like temple site of Jebel (or Gebel) Barkal, about 400km north of Khartoum. In Roche's version, Jebel Barkal "has been known about since 1949, but there never seems to be either funds or the interest to do the job". This is confusing, since major excavation at the site began in 1916, when George A. Reisner led an expedition mounted by Harvard University and the Boston Museum of Fine Arts; since the 1970s the University of Rome has been excavating there continuously,

---

* In the 19th century Heinrich Schliemann (1822–1890) made a very similar supposition about the writings of Homer and as a result located Troy.

and from the 1980s these efforts were joined by another
Boston Museum of Fine Arts project. It may be that Roche
was referring to the fact that some of the site's 13 (at least)
temples are not accessible to excavation because still
regarded as sacred by the locals. This is rather a stretch,
though, from "there never seems to be either funds or the
interest to do the job".

It's easy to come up with all sorts of theories about the
Pyramids from the comfort of one's armchair, a little more
difficult to match those theories up to the real-life rocky
objects. Confrontation with the real Pyramids is of course
no guaranteed cure for unorthodox hypotheses – Piazzi
Smyth's were actually *generated* by his on-site researches –
but it can have a salutary effect. Take the case of Mark
Lehner, a US worker who trained to be an archaeologist
with the specific intent of proving Cayce's revised chronology of ancient Egypt, and who in 1973, while still a student
at the University of Cairo, published a monograph called
The Egyptian Heritage which did its very best to marry
Caycean with orthodox archaeology. It's not long after this
point that Lehner rather drops out of the Caycean
hagiographies, perhaps largely because, working with such
icons of egyptology as Zahi Hawass (b1947), he discovered
the excitements of genuine archaeology and disowned his
Caycean views.

Curiously enough, for a while Lehner was under attack
by Graham Hancock and Robert Bauval for his earlier
Caycean associations, which the two unorthodox theorists
found sinister. Eventually, however, they magnanimously –
and very publicly – forgave him his sins and declared how
they hoped to able to work in future in an atmosphere of
mutual respect, even if disagreeing on various egyptological
matters. No more sleepless nights of worry, then, for one of
the world's foremost egyptologists.

Lehner's story in a way echoed that of Flinders Petrie
(1853–1942), one of the greatest egyptologists of all time.
Petrie first became involved with the Pyramids because his
father, a chemical engineer, swallowed Piazzi Smyth's wild
numerological notions hook, line and sinker, and determined to make ever more accurate measurements of the
Pyramids on-site. In due course the father became too old

Drawing by Flinders Petrie of the Pyramids of Giza. Initially travelling to Egypt in the belief he could prove Piazzi Smyth's hypotheses, Petrie instead became one of the greatest Egyptologists of all time

to travel to Egypt, and sent young Flinders on his own. In a sense, young Flinders never came back. His work surveying the Pyramids demonstrated conclusively, of course, that Smyth's hypotheses were nothing but self-deluding fantasy.

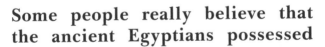

Some people really believe that the ancient Egyptians possessed supertechnology.

Much ink has been devoted to a group of supposedly mysterious inscriptions in the Temple of Seti at Abydos, in particular to one that bears a superficial resemblance to an Apache helicopter. It has been known for decades that this particular inscription is a badly executed palimpsest. Rameses the Great (1304–1237BC) had a habit of erasing the name of his father, Seti I (reigned c1290–1279BC), from monuments and superimposing his own, and this is what happened here. The job was sloppily done, and a certain amount of later wear has revealed a little of the underlying, imperfectly erased marking. That the result should look like a helicopter is of course coincidental . . . or is it?

Plenty of unorthodox archaeologists are quick to claim it as a depiction of exactly what they think it is. One such is Richard C. Hoagland (b1945), the theorist probably best known for his championing of the Face on Mars as an example of the architecture of a long-dead civilization. It's his contention that the ancient Egyptians were descendants of Martians who'd settled in that sandy region because it reminded them of home. Other theorists pooh-pooh this notion on the grounds that a nearby inscription is, they claim, a representation of a submarine. All very well, they point out, for Martians to have invented helicopters, but Mars has no seas and so the idea of a Martian submarine is obviously ridiculous. Hm. What they don't mention is that it'd be a real difficulty getting a helicopter to work in the parlously thin atmosphere of Mars.

A *Pravda* article dated November 11 2005, "Ancient Egyptians Used Helicopters and Airplanes for Battles?", has been widely quoted in various internet forums as a source for further information about the inscriptions. It cites various unorthodox theorists, like "well known Egyptologist" Alan F. Alford (b1961), author of such works as *Gods of the New Millennium* (1997), who argues that the Great Pyramid and the larger of its two companions at Giza were built millennia before the other Egyptian pyramids, the Khufu concerned being a far earlier king of the same name. I've been unable to track down two of the other experts cited in the *Pravda* piece. One is "famous Egyptologist" Bruce Rowles, who

> gives another interesting hypothesis about the origins of the strange hieroglyphs. He says there were no interplanetary expeditions from other star systems to this planet in the old times. He supposes that Egyptian pagan priests knew many of nature's secrets. It is a proven fact that 3,000 years ago Egyptians made the first batteries and generated electricity. Bruce Rowles also supposes that pagan priests in Ancient Egypt could look to the future where they quite probably saw battle helicopters, aircrafts and submarines.

The other authority I was unable to trace is William Deutch:

> Popular scientific literature says that . . . Tutankhamen died 3,300 years ago as a result of an air crash. Historian William Deutch . . . said that ancient Egyptians could go up to the clouds with balloons inflated with hot air and with primitive gliders. Such flights were available for the royal family and nobility only. Deutch says that many of the royal family in Ancient Egypt including Tutankhamen himself died with their legs broken and numerous wounds as if they tragically died as a result of an aircraft crash. . . . Deutch says that aeronautics first appeared in Egypt and then spread to the territories currently known as Tibet, India, Mexico, Turkey, China and Guatemala, in other words those territories where air flows could hold aircrafts in the skies.

It seems *Pravda* is upholding the grand tradition of scientific integrity it maintained throughout the Lysenko years. What next? Batboy?

Of all the writers who've pronounced on the Pyramids, perhaps none has spouted so much or such egregious nonsense as Erich von Däniken. Intent on demonstrating that the Pyramids must have been built using technology borrowed from visiting extraterrestrials, von Däniken seems to have drawn out of thin air such assertions as that the ancient Egyptians didn't have rope and that wood was rare because trees were scarce along the Nile. In reality, the Pyramid builders had plentiful rope – they often left the stuff lying around for archaeologists to find – and they used wood habitually in engineering and otherwise, trading with neighbouring countries to make up any shortfalls in domestic production.

Another of von Däniken's contentions is that the Egyptians mummified their dead so the remains could be reanimated when the extraterrestrials returned. A major hiccup for this hypothesis is that part of the process of mummification involves extracting the brain. This is not conducive to a successful resurrection. It might be countered that the Egyptians didn't realize the damage they were doing, because they believed the heart was the home of the soul and the brain of lesser significance. But wouldn't the highly sophisticated extraterrestrials have corrected their protégés' little anatomical misunderstanding?*

A sad irony is that, quite often, bodies buried by the Egyptians in the sand have been preserved better than those subjected to mummification. The dry sand effectively desiccated the corpses. It's not *quite* that returning extraterrestrials would merely have had to add boiling water and stir, but . . .

Another set of "mysteries" fulminated upon by von Däniken and his ilk concerns the ability of the Egyptians, using only the technology they had, to construct such mighty edifices to the astonishing levels of exactitude evident: the north–south, east–west alignments of the faces of the pyramids are precise to within a matter of a very few minutes of arc; the greatest difference between the lengths of the different sides of the Great Pyramid is a mere 20cm; the individual blocks may not always fit together quite so tightly that you'd be unable to put a piece of paper into the crack but the precision is nevertheless highly impressive. And how would the ancient Egyptians have been able to carve those stone blocks without the use of steel chisels?

Although in some cases we don't know exactly how the Egyptians performed specific tasks involved in the construction, that knowledge isn't needed in order to rebut the von Däniken-style arguments that only through reliance on the expertise of visiting extraterrestrials or Atlantean survivors could the Egyptians have coped. The fallacy in such propositions is an endemic and perhaps wilful failure to recognize that, while the Egyptians were more ignorant than we are and lacked much we take for granted, they were every bit as smart as we are. If *we* can work out how to build pyramids using solely the technology and scientific knowledge available to the ancients, we can be pretty damn' sure they could do likewise. And we *can* work out how to do it – usually with a choice of several different ways. Perhaps the Egyptians

---

* To be fairer to von Däniken than von Däniken ever is to logic, it is just possible to postulate that, after the aliens had departed, the Egyptians over succeeding generations forgot a lot of what they'd been taught and, in cargo cultish fashion, imitated actions they'd seen the aliens perform, in the primitive belief that the imitation was somehow as valid as the real.

didn't raise the blocks using long earthen ramps, for example, but the fact that they *could* have done it thus is all that's needed to obviate the argument that they must have used antigravitic superthrusters imported from Aldebaran. We know the builders could have achieved the precise matching of the stones to each other by setting them into approximate position while they were still rough-cut and then doing the finer adjustments *in situ*. And we know the stone and copper chisels the Egyptians possessed were perfectly adequate for carving the limestone of which the Pyramids were built.

In fact, there are some problems with the contention that the Egyptians used ramps to raise the stones for the Pyramids – such as that no ramps are mentioned anywhere in the contemporary records. Accordingly, the Australian mechanical engineer Paul Hai (b1952) proposed that instead the Egyptians, while they did not have the wheel, did have a device called the pinion pulley. He expounded his theory in the self-published book *Raising Stone I* (2007):

> This ancient pulley walks up (or down) steps in a similar way to a three wheel step-trolley and I firmly believe that this is how the ancient Egyptians built their Pyramids. The Giza Pyramids have steps which I have termed "racks" and there are four "racks" in a square based Pyramid. The wooden ancient Egyptian Pinion-Pulleys made positive engagements with the Pyramid's stone "racks" carrying a stone block each, rotating as they were being hoisted with ropes. No ramps were required as the Pyramid itself was used.
>
> This ancient pulley has a mechanical advantage of 2.8, thus is a simple machine and proves the Greek historian, Herodotus, to be absolutely correct as he recorded wooden machines made of short wooden planks were used to raise the blocks of stone.
>
> These planks only needed to be the side length of a Pyramid block which is about one metre and were easily carried also, as Herodotus also records. Well, Herodotus was only writing what Egyptian Priests told him, and Egyptian Priests recorded history as part of their duties.

There's another possibility, first advanced by the French chemist Joseph Davidovits (b1935) in the late 1970s and emphatically supported by the researches of Gilles Hug of

the French National Aerospace Research Agency and Adrish
Ganguly and Michel Barsoum of Drexel University,
Philadelphia, as presented in their 2006 paper
"Microstructural Evidence of Reconstituted Limestone
Blocks in the Great Pyramids of Egypt" in *Journal of the
American Ceramic Society*. They used primarily scanning and
transmission electron microscopy to compare samples from
the Pyramids with limestone from nearby quarries and
came to the conclusion that, while the stones of the lower
parts of the Pyramids were rocks cut to shape and hauled
into place, as per the standard model, those of the upper
levels were in effect made of concrete, cast *in situ*: that soft
limestone was dissolved in pools to make a watery slurry,
into which were mixed lime and salt before further
evaporation yielded a clay that could be carried in sacks to
the site, packed into wooden moulds, and allowed to
harden. Aside from the chemistry, a further point is that
(reportedly) the stones on the Pyramids' upper layers tend
to be denser at their bottoms than their tops, consistent with
the possibility that the hardened "concrete" blocks contain
trapped air bubbles in their upper regions, just like cement
blocks are prone to do.

The hypothesis has been resoundingly rejected by Zahi
Hawass, who as Egypt's Secretary General of Antiquities was
also interested to know where the scientists had got the
material they'd tested: "We certainly never gave permission
for anyone to take samples." Elsewhere he was quoted as
saying: "It's highly stupid. The pyramids are made from
quarried blocks of solid limestone. To suggest otherwise is
idiotic and insulting." As far as I can work it out, the reason
for the hostility of Hawass and other egyptologists is that
the scientists involved – including not just Hug, Ganguly
and Barsoum but also a number of others who've checked
their researches – are mere materials scientists, not archaeologists who, when working on digs, develop a sort of sixth
sense about these things. Well, yes. As Colin Nickerson
wrote in the *Boston Globe* for April 22 2008:

> Barsoum, a native of Egypt, says he was unprepared for the
> onslaught of angry criticism that greeted peer-reviewed
> research [i.e., the *Journal of the American Ceramic Society*

paper] . . . "You would have thought I claimed the pyramids
were carved by lasers," Barsoum said.

What then of the "mystery" of the motives for the Pyramids'
construction? For long enough it was assumed they were
tombs,* but it's very true that not a single body, mummified
or otherwise, of a deceased pharaoh has ever been found
inside any of them.† It's perfectly feasible they were tombs
but looted at some stage during the millennia; the practice
of graverobbing was already, we know, well established by
the time the Pyramids were built, and has been going on
ever since. Even so, it seems strange the putative robbers
could have effected such a comprehensive clean sweep.

An attractive suggestion was offered by the physicist
Kurt Mendelssohn (1906–1980) in his book *The Riddle of the
Pyramids* (1974): that what was important to the Egyptians
was not the completed monument itself but *the building of it*.
There have been plenty of examples in history of societies
undertaking massive public-works projects for no real
reason other than the necessity to hold themselves together.
It might well seem this could hardly explain such a colossal
expenditure of the nation's manpower and wealth, but we
must remember it was being done at the behest of a god –
for that is what the pharaoh was to his people. By way of
comparison, in the service of their god the people of Paris

*This wouldn't negate the notion that they had astronomical
purposes as well.

† This would seem, by the way, to put in an odd position the arguments of those, quieter today but vociferous during the 1970s and
1980s in the wake of the 1973 publication of Lyall Watson's bestselling book *Supernature*, who insisted the Egyptians built the
Pyramids because of the miraculous powers the very shape of a
pyramid has to preserve objects placed within. If the Egyptians
didn't think this after all, then where did that particular conceit
come from? Could it possibly have come from the book *Pyramid
Power: The Millennium Science* (1973), written and privately
published by G. Pat Flanagan (b1944) of Glendale, California,
whose business is selling pyramidal knick-knacks and running
expensive seminars promoting the notion of pyramid power?
Perish the thought!

took from 1160 to 1345 – nearly two centuries – to build Notre Dame Cathedral, and other medieval cathedrals took even longer.

Many theorists have attempted to make a connection between the Pyramids of Egypt and those of Central and South America, the earliest of which are the Temple of the Sun and the Temple of the Moon at Teotihuacan, Mexico. We don't need to detail – and won't – all the predictable theories which say the similarities of form arose because both sets of ancient builders were following the instructions of, or imitating, space people or Atlanteans, or simply that the builders were *themselves* space people or Atlanteans. Alternatively, unknown to archaeology, the Egyptians might have crossed the Atlantic – although, as Barry Williams put it in a 1988 article ("Pyramids, Pyramyths & Pyramidiots") in *The Skeptic*, "It would seem to be highly implausible that Egyptians, at the final stages of their long history, would venture halfway around the globe and then teach the natives a technology that they themselves had abandoned nearly two millennia earlier."

Besides, the similarities of form are not that close: the Egyptian and American structures are both pyramidal, and that's about it. The angles, materials, details and functions differ. In essence, the pyramid can be seen as a natural development of the earthen mound, and the pyramid form as one that's inevitable when cultures that have no knowledge of the principles of the arch want to build large structures that won't readily fall over.

As the 20th century drew to its close, the Egyptian Government began to display nerves about the potential influx of New Agers and worse. According to an *Agence France-Presse* report on December 10 1999:

> At least two spiritual tour operators are advertising millennium trips to the pyramids, billing guest speakers such as David Icke . . .

Mr Icke and others are warning that former President [George H.W.] Bush will summon oppressive evil forces at a black mass in a dank stone burial chamber deep inside the great Cheops pyramid at midnight on Dec. 31.

According to [Icke's] Web site, Mr Bush and the British royal family are key members of the world "Illuminati elite" of human–reptile hybrids whose rituals are designed to tap into fourth dimensional energy forces and deprive ordinary human mortals of their consciousness. . . .

Police presence will be stepped up both at the pyramids and in the nearby desert where 50,000 partygoers will be channeled into an all-night, end-of-year spectacle starring multimedia artist Jean Michel Jarre. . . .

The notorious bi-weekly newspaper *Al-Shaab* . . . has accused Mr Jarre of conspiring to impose "Zionist Freemasonry" on Egyptian civilization. . . .

Texe Marrs, a retired Air Force officer, meanwhile suggests on his Web site that Mr Jarre's concert, which will be staged in the desert around half a mile away from the pyramids themselves, is a diversion from the Illuminati rituals of "consummate evil."

"The grotesque ceremony, these men believe, will culminate in a visit by their glowing Masonic god of light and magic, Lucifer himself, at exactly the stroke of midnight, December 31, 1999," he writes.

Some people really believe other ancient monuments show the presence of aliens or supertechnology or both.

An especial favourite of the bogus theorists is the Nazca Lines, found on a high arid plateau that lies between the towns of Nazca and Palpa, Peru. Most of the hundreds of designs created by removing the red stones of the surface to reveal the pale, chalky soil beneath are simply lines or elementary geometrical shapes, but a few dozen are more ambitious, being stylized depictions of humans or animals, and can be over 200m across. Radiocarbon dating and similarities of artistic style independently indicate the creators of the markings were almost certainly the Nazca Indians, who dominated the region between 200BC and 700AD. Since the images are clearly visible only from the sky, there has been considerable speculation about the method used to lay out the designs.

The purpose of the Nazcas in creating the enormous artefacts is, of course, something we may never be able to discover, although Gerald Hawkins (1928–2003) and others, using computer analyses, demonstrated the motive could have been astronomical, as there are far more alignments related to the solstices and other regular celestial events than mere chance would predict. The motive is more usually assumed to be religious, although the nature of the deities is open to debate, running from pantheist to extraterrestrial – what better way, too many theorists urge, to explain how the Nazcas could lay out the designs than by offering them the help of aliens, either technical guidance or through use of flying saucers for aerial progress checks? Why the aliens might want this primitive people to construct pictures for the aliens themselves to look at from on high is problematic, since presumably any technological culture sufficiently advanced to be able to travel between the stars would also have invented television. An alternative version, that the Nazcas constructed the images after the

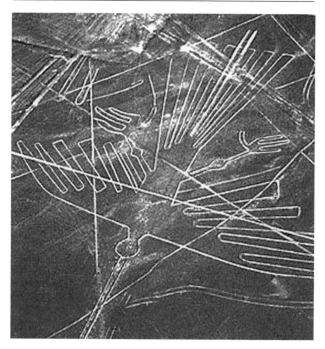

The great condor marking on the Nazca plateau

aliens had departed in a woebegone effort to entice the
skygods back, assumes the Nazcas were capable of laying
out the images without any aerial assistance and thus obviates the main reason for invoking aliens in the first place.

Wilder theorists insist the designs, while they have their
decorative aspects, are more importantly of a functional
nature – they're the runways, parking bays, etc., of an
ancient spaceport. There's no evidence whatsoever of the
spilled fuel or combusted ground areas you'd expect if the
aliens were to have conducted their takeoffs and landings
using any form of propulsion that might necessitate their
putting their spaceport on a remote, bleak plateau rather
than somewhere convenient to their HQ – wherever said
HQ might be.

That the Nazcas might have been capable of flight themselves is a by contrast perfectly respectable notion that has been advanced by, notably, Jim Woodman, who in the book *Nazca: Journey to the Sun* (1977; vt *Nazca: The Flight of Condor I*) described how he and Julian Nott constructed, using only technology available to the Nazcas, a hot air balloon in which he was able to fly above the inscribed plateau. Of course, proving something *could have been* done is not the same as proving it *was* done, and few archaeologists support his hypothesis. If nothing else, had the Nazcas discovered the secret of the hot air balloon, surely it would have spread to other South American cultures. There's no evidence that it did, or indeed of the Nazcas having had such a device.

In 1982 Joe Nickell (b1944) and a team of family members set out to reproduce one of the larger Nazca markings (the condor, 134m long) in an area of landfill near West Liberty, Kentucky. Rather than use sophisticated surveying techniques or the guidance of extraterrestrials floating overhead in flying saucers, the Nickell team planned out their figure thus:

> The method we chose was quite simple: We would establish a center line and locate points on the drawing by plotting their coordinates. That is, on the small drawing we would measure along the center line from one end (the bird's beak) to a point on the line directly opposite the point to be plotted (say a wing tip). Then we would measure the distance from the center line to the desired point. A given number of units on the small drawing would require the same number of units – larger units – on the large drawing.[*]

The results were surprisingly good. The condor is one of those Nazca markings where the 6th-century creators seem to have erred in their technique, giving the bird asymmetrical wings and feet (so much for those clever airborne aliens, hm?); the Nickell team recaptured this aspect perfectly. Again, such an experiment doesn't prove the Nazcas *did* use

[*] Nickell, Joe: "The Nazca Drawings Revisited: Creation of a Full-Sized Duplicate", *Skeptical Inquirer*, Spring 1983

this particular method, but it seems far more probable they used something like it than that they were guided by extra-terrestrials or even by humans sailing aloft in hot air balloons.

Ica is a place in Peru, and in caves and stream beds near it have supposedly been found thousands of stones bearing prehistoric carvings that have the potential to turn the worlds of archaeology and palaeontology on their respective heads. The carvings show items such as aerial maps, humans interacting with dinosaurs and other extinct animals, and advanced technological devices such as tele-scopes and life-support machines. The stones were the subject of two BBC television documentaries, in 1977 (gullible) and 1996 (rational).

The stones owe their fame to the fact that a physician called Javier Cabrera was given one as a gift in 1961. He assumed the carving on this stone was of colossal antiquity because it depicted what he recognized as an extinct species of fish. He made it known he was in the market for more of these intriguing objects, which was a foolish thing to do. Several hundred stones were sold to him by a pair called Carlos and Pablo Soldi, who'd experienced difficulty in sell-ing similar artefacts to professional archaeologists, and thousands of them by a peasant farmer called Basilio Uschuya.*

Much later, in the 1990s, the Peruvian authorities arrested Uschuya for selling archaeological discoveries, a crime in that country, and he promptly admitted he and his wife had faked Ica stones; he was released and allowed to continue his trade, because the law forbids only the selling of *genuine* archaeological artefacts. Naturally, apologists such as Sylvia Browne, in her *Secrets & Mysteries of the World*, say Uschuya lied about having manufactured the objects in order to get himself out of trouble; they don't explain the

---

* Cabrera's experience in this respect echoed that of Johann Beringer (1667–1740), who was anxious to show fossils were arte-facts planted by God. Sure enough, people in the region soon found plenty of evidentiary examples – fossils showing Christian symbols, etc. – which they sold to him by the hundreds and thou-sands.

scene in the 1977 documentary where he created and
"aged" an Ica stone for the cameras using a dentistry drill
for the engraving and baking in cow dung for the celebrated "patina of age" the carvings exhibit.

On the basis of the maps on some of the stones,
Cabrera believed the carvings dated from 13 million years
ago, which is long before any accepted date for the appearance of *Homo sapiens*, let alone human civilization. Browne
argues that the date's about right for the carvers to have
been Atlanteans, for by then this race had arrived from the
stars to settle the doomed continent, and they looked much
like humans except several feet taller. Of course, her suggestion doesn't help with the problem of the depicted
dinosaurs being around nearly 50 million years after their
extinction.

Cabrera, too, invoked extraterrestrials in his attempt to
explain the artefacts. All told, he accumulated well over
10,000 of these, and he attempted to put them in some sort
of chronological order, so they'd offer a narrative of the
culture that had produced them. This he accomplished to
his own satisfaction, coining the term Gliptolithic Man to
describe this lost race. Gliptolithic Man was not native to
this planet but had arrived to colonize it from one of the
stars of the Pleiades. Later on, not long before their departure, the Gliptos used genetic engineering to create, as a
sort of farewell present, a new intelligent species here on
earth from a native animal whose form chanced slightly to
resemble their own. The reason for their departure was that
the earth was becoming dangerously unstable, largely as a
result of the Gliptos' activities. You see, although this is
unrecognized by modern geology, the earth's surface of 13
million years ago or so consisted largely of land, with only
about 20% open water. The atmosphere was unbearably hot,
presumably due to some sort of greenhouse effect. (Perhaps
all the missing surface water was present in the form of
atmospheric vapour? Who knows?) The Gliptos attempted
to ameliorate the environment through manipulating
Nature's biological cycles, but their efforts triggered a
massive, worldwide tectonic upheaval* that eventually

* No, I don't understand this bit either.

resulted in something much closer to the world we know today. By then, though, the Gliptos were long gone, having fled back to their Pleiadian birthplace.

In 1998 a Spanish investigator called Vicente Paris announced he'd demonstrated through microphotographic analysis that the carvings had been produced using modern abrasives. Far earlier, most people who'd seen the stones had been confident of their modernity because the carvings showed no sign of the gross erosion they'd have suffered had they indeed been millions of years old.

A similar observation discounts the validity of the so-called Acámbaro figures, a set of ceramic figurines discovered near Acámbaro in Mexico and depicting such items as dinosaurs. The first was spotted in 1944 by one Waldemar Julsrud when out riding. He told the locals that he'd be interested in purchasing any further examples they might unearth during their rustic pursuits . . . and soon the ceramics started pouring in, to the tune of, eventually, some 32,000. An investigation by the distinguished US archaeologist Charles C. Di Peso (1920–1982) demonstrated conclusively the figurines were fakes; he published his results in a 1953 paper in the scholarly journal *American Antiquity*. This did not stop Charles Hapgood (see page 135) from writing his supportive *Mystery in Acámbaro: An Account of the Ceramic Collection of the late Waldemar Julsrud in Acámbaro, Gto., Mexico* (1973).

But are we being a tad unambitious? Should we limit ourselves to edifices constructed on our own planet? What about the celebrated Face on Mars? Much touted by Richard Hoagland and others for decades as obviously an item of Martian architecture, this has been revealed by more recent NASA photographs to be a perfectly ordinary geological formation that chances to resemble a face when seen from a certain angle. Even so, do we have evidence of off-planet master-builders elsewhere?

According to George H. Leonard (b1921), in *Somebody Else is on the Moon* (1976), two or more species of aliens are currently performing mighty feats of engineering all over the moon; an especially conservationist measure is the sewing up of great gashes in the lunar surface. Leonard claims the presence of his aliens explains all sorts of lunar conundrums – such as the bright "rays" which spread out from many craters. Most planetary scientists think these are ejecta thrown out as the crater-forming meteorites impacted, but the true explanation must surely be this: Many craters are floored with whitish dust. Spacecraft travelling to another part of the lunar surface from such a crater naturally start out with a fair amount of the dust clinging to their underbellies. As they travel, this dust falls off. Over the millennia . . . need we elaborate?

Leonard's text is backed up by NASA photographs which, squint at them as this reader might, seem not to show any of the details he says can be seen there.* Luckily we have his sketches of those invisible details to help us, and they certainly seem pretty impressive. There are engineering rigs of enormous size, some taking the form of the letter "t" and others, more organic in appearance, that of the letter "X" – they resemble, as Leonard observes, two worms laid over each other at right angles. There are equally massive generators and gear systems, plus enigmatic shapes that could be vehicles but equally well, Leonard observes, could be gigantic alien yurts. There is more, much more, to indicate the presence and frenetic engineering activities of these aliens.

Leonard uses a few narrative techniques that might not be acceptable were he to have published his findings in the form of, say, a paper in *Nature*. In one instance, lacking any evidence that *Apollo* astronauts actually saw a particular lunar feature, he supplies their dialogue anyway: "We know from the tapes how the astronauts reacted to other phenomena. Discovery of this huge object might have gone like this . . ." And his primary Deep Throat, the ex-NASA engineer who gave him pointers and tips as he was conducting his investigation, is made known to us only as a pseudonym, Sam Wittcomb. Wittcomb displays an inexplicable – and quite infuriating – reluctance to tell Leonard anything outright, instead, Yoda-like, giving him cryptic clues and smug recommendations.

---

* In fact, NASA has images of the relevant areas taken much closer to the lunar surface, and therefore showing much more detail – detail that would undermine Leonard's beliefs. He therefore ignores them.

That the moon might be an alien spaceship, damaged somewhere between the stars and parked in earth orbit for a few hundred thousand years' worth of repairs, Leonard regards as merely a hypothesis, although it's a hypothesis to which he returns several times – and which he extends, speculating that those moons of other planets that move in retrograde orbits might also be parked alien spaceships, as might be, in a suggestion not original to him, the two small moons of Mars.

Leonard has a way with words. Galileo is a "feisty pioneer". Some phrases are impenetrable: "Out of a little more than two and a half tons of iron, a ton of oxygen can be extracted during the reduction process." Overall, though, he satisfies himself that

> The prime reason for the United States' launching an expensive Moon program (and sending spacecraft to Mars and beyond) was the recognition at official levels that the Moon (and perhaps Mars) is occupied by intelligent extraterrestrials who have a mission that does not include dialogue with us and may even be inimical to our long-range welfare.

Some people really believe most of the Dark Age never happened. The term "Dark Age" came about because, in Western Europe for a period extending roughly from the 5th century to AD1000, there seemed a paucity of historical records; the obvious conclusion was that this had been a period of cultural stagnation when barbarians had ruled and slaughtered. More recently historians have rejected this assumption; while they wouldn't exactly describe the period as marked by cultural glory, they nonetheless see it as far more benign than their predecessors did. Accordingly the period is now more acceptably termed the Early Middle Ages.

But what if in fact it never existed at all? Thus posits the Phantom Time hypothesis put forward by the German systems analyst Heribert Illig (b1947) in books like *Das Erfundene Mittelalter: Die Grösste Zeitfälschung der Geschichte*

(1996) and *Wer Hat an der Uhr Gedreht? Wie 300 Jahre Geschichte Erfunden Wurden* (1999). Resolutely multidisciplinary as he builds his case, Illig makes claims that have not gone down well in the historical community – most notably that Charlemagne (747–814) was invented for cultural and nationalist reasons by later writers, much as Geoffrey of Monmouth (*c*1100–1154) invented, or at best constructed on the basis of wisps of oral legend, great slabs of English pseudohistory. It is of course far from outrageous to suggest that earlier historians made things up at the bidding of kings or whim – figures like Thucydides (*c*460BC–*c*400BC) and Plutarch (*c*46–*c*120) did exactly this, and it's only recently that the "science" of history has emerged with its crazy notion that historians should attempt to portray *what really happened*. Even in the 21st century we've had blatant examples of large-scale attempts to rewrite recent history, such as the later assertion by the Bush Administration that the US invaded Iraq in 2003 because the Iraqi Government had refused to admit UN inspectors to check for weapons of mass destruction.* One shudders to think what might pass for history in, say, North Korea.

It seems that what triggered Illig's suspicions was an archaeological conference held in Munich in 1986 whose subject was the widespread forgery of documents in the Middle Ages. Simply because a document has been forged doesn't necessarily mean its contents are wholly or even partially untrue; for example, a scribe might set down well established facts and then backdate his finished document to enhance its perceived authenticity. Other forgeries need not deceive the historian: a monk might forge his abbey's title deeds because the originals were lost and a dispute over ownership has now come up.† But some medieval forgeries

---

* This travesty is still being repeated. The reality is that the UN inspection team, under the leadership of Hans Blix (b1928), fled Iraq days before the invasion in order to get out of the line of fire.

† It can also be argued that a forged document is actually evidence for the existence of its subject: had the content been *entirely* invented, it would have likely been rejected as a lie by contemporaries.

are far more misleading, and Illig and his supporters focus particularly on a number that were produced by the Roman Catholic Church before about 600 that *only after about 900* would come in handy for the Church. How could those ecclesiastical forgers have been so prescient? Could it be, rather, that there's an error in our chronology of European history, such that there's a jump in the counting between the years 614 and 911? Could the "history" of those three centuries be a fiction, concocted later for political or theological reasons or simply in an attempt to make sense of the chronology?

One thing we should remember here is that the BC/AD system of counting the years, so familiar to us today, was not in fact invented until *c*525, by the monk Dionysius Exiguus (d556), and didn't really take widespread hold until some centuries later when the imminence of the millennium began to assume importance. For a long time in Europe there survived the tradition of counting years from the supposed foundation date of Rome (reckoned as 753BC by our system). More common was the practice of dating years according to the reigns of monarchs or the incumbencies of popes: it was far more relevant that this was the seventh year of John III than to know which century you were living in. The vast majority of people would have been ignorant of the fact had they fallen asleep in 614 and woken up in 911, and, even if they'd known about it, wouldn't have cared because that wasn't the system of counting they were using anyway.

His suspicions aroused by knowledge of the medieval forgeries, Illig turned his attentions to the introduction of the Gregorian calendar in 1582 by Pope Gregory XIII. The effects of the approximation of the Julian calendar, which Julius Caesar had introduced 1627 years earlier, in 45BC, were by now becoming too large to ignore. In particular, in order to get things back into synch there had to be an adjustment of 10 days. Illig did his math and pointed out that, after a period of 1627 years, the discrepancy should have been 13 days, not 10. A difference of 10 days implied that the Julian calendar had been instituted a mere 1257 years before. And this would indicate that the Julian calendar was introduced in AD325 – a stark impossibility, according to accepted chronology, because Caesar would have been dead for over three centuries by then!

There's nothing wrong with Illig's math. The adjustment that had to be made with the introduction of the Gregorian calendar does indeed lead one to the date of AD325. That was the year of the 1st Council of Nicaea, one of whose tasks was the establishment of a means of calculating Easter. The date of the vernal equinox was an important element of such a calculation; in AD325 it was March 20. By AD1582, because of the inaccuracy of the Julian calendar, the vernal equinox was now falling on March 10 – hence the necessary 10-day discontinuity when the replacement calendar was brought in.

Illig and his followers, confronted by this, claim there's no evidence the Council of Nicaea did any such thing, so there.

There are other evidences adduced by Illig and the stalwart band of scholars who support him. One of these, Hans-Ulrich Niemitz (b1946), identified numerous of these in his paper "Did the Middle Ages Really Exist?" (1995). For example:

> . . . a gap in the history of building in Constantinople (558AD–908AD); a gap in the doctrine of faith, especially the gap in the evolution of theory and meaning of purgatory (600AD until ca.1100 . . .); a 300-year-long reluctant introduction of farming techniques (three-acre system, horse with cummet) and of war techniques (stirrup) . . .; a gap in the mosaic art (565AD–1018AD); a repeated beginning of the German orthography, etc., etc. . . . The puzzles of historiography led the way, pointing out again and again the "gap" which we soon termed "phantom time".

Surely, though, the scientific dating techniques used in archaeology could sort the problem out once and for all? The two most important are carbon-14 dating and dendrochronology. In fact the former isn't suited to timescales as short as the ones we're talking about, so we're left with dendrochronology, the dating of historical events through matching them with tree rings.* And here we hit a

---

* The size of a tree's annual ring depends upon the environment – essentially, the climate – in the year during which the ring formed.

difficulty that could be regarded as a point in favour of Illig's thesis: there are far too few pieces of usable wood extant from the period in question to allow any accurate dendrochronological dating. Could this be because – cue drumroll – those few examples have been misinterpreted and really there are *no examples at all* from the period, for the very good reason that the period never existed?

Assuming the will to forge the existence of a three-century period, surely the enterprise would be impossibly huge? It wouldn't be a matter of just writing a narrative which all subsequent readers would accept as a true history. No, great numbers of documents would have to be faked; further, many *existing* documents would have to be re-edited – or rewritten, or in some cases merely destroyed – in order to take account of the new version of things. An entire culture would have to be complicit in the conspiracy.

Yet this is not as impossible as it might seem. In the 9th and 10th centuries, in consequence of the introduction of the newfangled minuscule script to replace the more cumbersome majuscule, the entirety of Byzantine literature was transcribed over a period of no more than two or three generations. The original majuscule documents are lost to us. There's no reason to believe Byzantine history was extensively rewritten as part of the exercise in order to pander to the relevant emperors; but, inarguably, it *could* have been, and today we'd be none the wiser. It's perfectly conceivable there may have been exercises of mass rewriting of which we know nothing.

But what motivation could there have been for changing the numbering of the years, and by such a large jump as nearly 300? Illig's primary suspect appears to be the Holy Roman Emperor Otto III (980–1002). Otto was a keen patron of literature, scholarship and the arts. Is it feasible he was born not near the end of the 10th century but instead near the end of the 6th? Might he, for religious

---

If you observe a sequence of rings in one tree you can hope to identify the same pattern in other trees. By repeating this process with progressively older and older trees or wooden artefacts (e.g., roof beams), one can attain quite precise dates for events even well back before the time of Christ.

reasons, have decided he wanted to be the one on the throne of the Holy Roman Empire when the millennium came? With the aid of his mentor Gerbert of Aurillac (*c*940–1003), who was Pope Sylvester II from 999, couldn't Otto have conjured up a scheme to renumber the years? Indeed, it's possible that, if Otto were flourishing around AD700, no one might have been too clear *what* the date was by the BC/AD count, and so Otto might simply have been introducing a bit of certainty into the dating system.

The second part of this scenario envisages historians gazing back at an apparent three-century tract of time that had appeared as if from nowhere, seeing it had nothing in it, and setting to work with a will to fill it with fictitious events and personages – including Charlemagne and his many exploits.

To be sure, the picture that emerges of Charlemagne from such sources as the 11th-century epic *La Chanson de Roland* is so enshrouded with legend as to resemble a figure whose historicity most certainly *is* in doubt, King Arthur; but this isn't sufficient to say with any confidence that he didn't exist: there seem far too many evidences of his life and realistic achievements. Of course, a coordinated effort by many scribes to rewrite history could have generated a plausible biography for Charlemagne, with all the fanciful embellishments added later, but who could have devised such a coordinated effort, and how in the world could it have been organized in an era when communications between one centre of learning and another could be hazardous and agonizingly slow?

Is it not much more likely that Charlemagne lived, and that his biography was much as the historians have worked it out to be, even if some of the details may be in error? There are countless examples in history, even very recent history, of the lives of perfectly ordinary mortals being mythologized enthusiastically after their deaths: just think of Elvis Presley. There's also the tendency to embellish biographies for reasons of emotional bias and political expediency: Charlemagne was a mighty figure and a great hero, so naturally his story was rapidly girt with legend; conversely, the biographies of someone like Nero (37–68) were written by people who had good reason to loathe him so we really have no reason to think the picture we have of him is authentic. Furthermore, isn't Illig tending to ignore the fact that, in history, by and large hard facts become fewer and more difficult to pin down the farther back in time you go? It may really be the case that there's not much to be found from the Dark Ages – or the Early Middle Ages – because not much was written down and most of what was has since been lost.

Some people really believe you could build a machine which, once started, would carry on running forever. Even a casual roam around relevant internet sites on the topic soon reveals that, when too many people discuss and promote alternative energy sources, they're not talking about solar panels or wind turbines. The dream of getting usable energy from nowhere has inspired inventors since at least the times of Classical Greece, long before the impossibility of the dream was realized: there can be no such thing as a perpetual-motion machine – at least, not without altering one of the fundamental laws of the observed universe, that in any system energy is conserved. In other words, you can't get more out of a machine than you put into it. In fact, you can't even get *the same* out of a machine as you put into it, because no machine is 100% efficient. Nevertheless, such putative devices have attracted investors numerous times over the centuries. And the investors, in turn, have attracted both misguided and bogus technologists . . .

The earliest European design of which we have a diagrammatic record is that proposed by the 13th-century French architectural artist **Villard de Honnecourt** (*fl*1225?). This shows one of the archetypal patterns for such devices. To the rim of a massive vertical wheel are attached a number of heavy mallets, each hinged to the wheel at the base of its handle. Once the wheel is set in motion, each

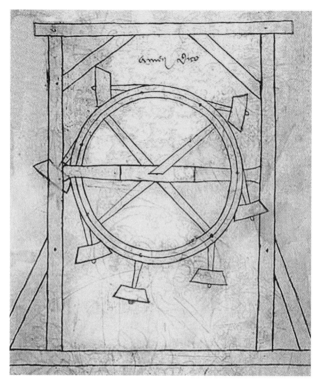

Villard de Honnecourt's perpetual motion machine

mallet flops over under the influence of gravity as it reaches the top of the cycle, and thereby adds impetus to the wheel's further rotation. The notion is that in this way gravity will keep the wheel rotating indefinitely. In reality, since as much work needs to be done raising the mallets to the top as is gained from their flopping over, only in an absolutely frictionless environment would the device work – but in such an environment a plain wheel would keep turning longer than this one. Even as a thought experiment, then, Villard's device is useless.

Nonetheless, the appeal of the basic design has ensnared many – among them **Leonardo da Vinci** (1452–1519). In Leonardo's sketchbooks there are doodles of a number of devices that are essentially variations on Villard's concept. Some show wheels whose spokes are partially submerged in water, as if in hope that buoyancy might add to the endeavour. Some hope. Leonardo concluded that at least this form of perpetual motion was a chimera:

> . . you might set yourself to prove that by equipping such a wheel with many balances, every part, however small, which turned over as the result of percussion would suddenly cause another balance to fall, and by this the wheel would stand in perpetual movement. But by this you would be deceiving your-self . . . As the attachment of the heavy body is farther from the centre of the wheel, the revolving movement of the wheel around its pivot will become more difficult, although the motive power may not vary.

Even earlier than Villard's version of the overbalanced wheel were some Arabic designs along the same lines, and they in turn had as precursors some designs recorded by an Indian writer called **Bhaskara** (*fl*1160), again based on the idea that heavy items appropriately attached to the rim of a wheel will keep it turning. In one of these designs, cylinders partly filled with mercury take the role of Villard's mallets. In another that uses the same principle, even though it looks superficially quite different, the wheel has hollow, appropriately curved spokes that are about half full of mercury. Leonardo's doodles contain something similar, although in his instance, rather than hollow spokes, the wheel has several curved chambers that have heavy marbles in them.

The quest for perpetual motion continued – the subti-tle of Arthur W.J.G. Ord-Hume's 1977 book *Perpetual Motion: The History of an Obsession* says it all. The Italian natural philosopher **Mark Antony Zimara** (*c*1460–1523 or 1532) in 1518 published an essay called *Directions for Constructing a Perpetual Motion Machine without the Use of Water or a Weight*. In it he described his invention, although admitting he'd never actually built it:

Construct a raised wheel . . . like the wheel of a windmill, and opposite to it two or more powerful bellows, so arranged that the wind will turn the wheel swiftly. Connect to the periphery of the wheel, or to its centre – whichever the builder thinks better – an instrument which will operate the bellows as the wheel itself turns . . . The wind which comes from the bellows and blows against the vanes of the wheel will cause the wheel to rotate, and the bellows themselves, powered by the rotating wheel, will blow perpetually. This, perchance, is not absurd, but is the starting point for investigating and discovering that sublime thing, perpetual motion . . .

In his *Novo Teatro di Machine et Edificii* (1607) the Italian mechanic **Vittorio Zonca** (1568–1602) presented a quite different form of device. Here there's an arrangement rather like that for a siphon – except the water-level of the input is lower than that for the output. As the water flows back down to the reservoir it turns a turbine.

For the output of a siphon to be higher than its input was something known as impossible since ancient times. While a siphon's tube is higher in places along its length than the reservoir, if the *output* is higher – or if the pipe is ruptured at a level higher than the reservoir's – nothing works. Water finds its own level or, if given the opportunity, flows downhill. Zonca seems to have tried a method of subverting this basic principle. In his day people knew nothing of atmospheric pressure, although they did know of the Archimedean notion that nature abhors a vacuum and could see its role in the functioning of a siphon. Zonca apparently thought what drove an orthodox siphon was the difference in weight between the column of water on the output side and the shorter column of water on the input side. In his device, therefore, he made the pipe on the output side very much broader – so that, even though shorter, it could bear a greater weight of water than its input-side counterpart.

In a 1618 folio the English alchemist **Robert Fludd** (1574–1637) illustrated and described numerous machines of his own devising, some intended as practicable and others as thought experiments. Water pumps were popular

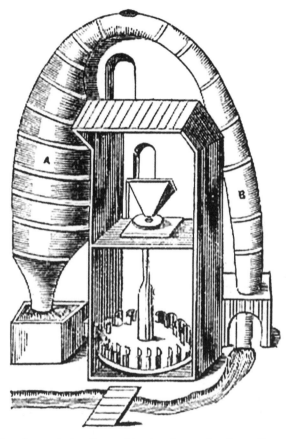

Zonca's perpetual motion machine, based on a misunderstanding
of the principles of the siphon

among these designs, and one is a perpetual motion
machine: a pump raises water to drive a wheel that in turn
powers the pump. However, Fludd makes it plain he knows
this device, which he attributes to "an Italian", could never
work in reality:

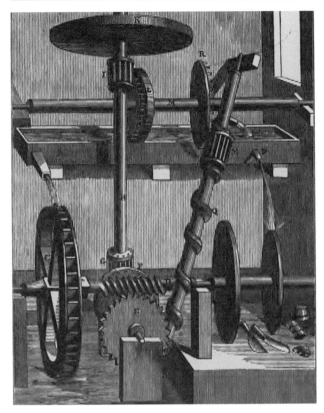

Although Fludd recognized others' perpetual motion machines
wouldn't work, that didn't stop him designing this mill . . .
which wouldn't work

It is unnecessary to point out that this appealing principle has
been tried several times, by people who often were completely
inexperienced in mechanical principles, and who did not see
the serious error which lies in these devices, and therefore
wasted effort, money and time on old and worthless ideas.

His stricture did not extend to the idea of perpetual motion
in general, though. He himself designed a mill that relied

on a turning water wheel driving an Archimedean screw that raised the water used to turn the wheel. Suspicious, the Bishop of Chester, John Wilkins (1614–1672), recounted in his book *Natural Magick* (1648) his attempt to build a working model of Fludd's device. Wilkins concluded it was an impossible machine.

Unfortunately the inventors of putative perpetual motion machines paid little attention to Wilkins's conclusion. A New York patent attorney declared ruefully in 1871 that it was a rare year if he didn't have at least one variant on Fludd's water mill submitted to him.

Long before the attorney made his complaint, the German automaton-maker, Jesuit priest and polymath **Athanasius Kircher** (1601–1680) – inventer of the ear trumpet and the megaphone – was another in the long line of those producing designs that worked (or didn't work) on the same principle as Fludd's. Kircher was interested in magnetism, too: he devised a magnetic clock and toyed with magnetically driven perpetual motion devices.

Indeed, magnetism must have seemed like a highly promising avenue to pursue, since magnetic force was a seemingly inexhaustible source of motive force. Gaspar Schott (1608–1666), in his book *Thaumaturgus Physicus, sive Magiae Universalis Naturae et Artis* (1657–9), describes another magnetic device, this time the brainchild in the 1560s or 1570s of the Jesuit priest **Johannes Tausnierus** (Johann Tausner). A powerful lodestone drew a series of metal balls up a ramp; as each ball neared the top it fell through a hole onto a lower ramp, down which it ran to join the reservoir of metal balls at the bottom waiting to be drawn up the upper ramp again. The impracticability is evident: if the lodestone were powerful enough to pull the balls up the ramp, very soon they'd all be stuck together in a clump at the ramp's top, as close to the lodestone as they could get. Nonetheless Bishop Wilkins, who'd been so admirably sceptical about Fludd's water mill, thought there might be some mileage in a design like Tausnierus's – even while admitting the Jesuit's machine itself was infeasible: "The bullet would not fall down through the hole but ascend to the stone."

Something similar to Villard's wheel was demonstrated

*c*1640 to King Charles I by Henry Somerset, **Marquis of Worcester** (1577–1646), and described by the latter as #56 of his *Century of the Names and Scantlings of Inventions* (written 1655, published 1663, rendered here from the 1813 edn ed John Buddle):

> To provide and make that all the weights of the descending side of a wheel shall be perpetually further from the centre than those of the mounting side, and yet equal in number and heft to the one side as the other. A most incredible thing, if not seen, but tried before the late king (of blessed memory) in the tower, by my directions, two extraordinary ambassadors accompanying his Majesty, and the duke of Richmond and duke Hamilton, with most of the court attending him. The wheel was 14 foot over, and 40 weights of 50 pounds a-piece. Sir William Balfore, then lieutenant of the tower, can justify it, with several others.—They all saw, that no sooner these great weights passed the diameter line of the lower side, but they hung a foot further from the centre,—nor no sooner passed the diameter-line of the upper side, but they hung a foot nearer. Be pleased to judge the consequence.

Whatever Worcester might say – and he's credited as an inventor of the fire engine and of a precursor of the steam engine, so cannot be lightly ignored – the "consequence" cannot have been much.* The civil engineer Henry Dircks (1806–1873), who in 1864 published an extensive analysis of the *Century*, was able to produce viable interpretations of the other 99 of Worcester's descriptions, but was at a loss in this instance:

> It is difficult to reconcile the statement he has here made, with the declaration on the title page, of his inventions having been "tried and perfected." . . .

---

* In 1770 Dr William Kendrick (d1779) claimed "the wheel was polite enough to turn about while his Majesty was present, [but] could not be prevailed upon to be so complaisant in his absence". We can't tell whether Kendrick learned this from some contemporary source now lost to us or whether he was just being waspish. Kendrick admitted no one had ever been able to get perpetual motion to work and that "the mathematicians" (whom he detested) had demonstrated it was impossible, yet still he sought its grail himself, citing past authorities to support his belief the achievement was practicable.

Dr Desaguliers, in a memoir, published by the Royal Society, vol. 31, 1720–21, quoting the foregoing article, ventures the reply: "Now the consequence of this, and such like machines [assuming them to be as above described,] is nothing less than a perpetual motion." Of course he does not admit even the possibility of such an arrangements of parts, he only allows that if that could be executed, the other would follow. But Desaguliers admitted too much, for it may easily be demonstrated that the conditions stated may be mechanically produced, without any resulting motion.

Dircks then demonstrates precisely why there'd have been no "resulting motion" before adding:

His notice of this exhibition was not written by the Marquis until 1655, from 14 to 17 years after its occurrence, and he may have then hesitated to say that it was not a success; but he may have persuaded himself that he was at last in possession of the secret that was at first wanting.* Besides, we are not to infer that the company described as being present had gone to the Tower purposely to see the Marquis's wheel; it being far more probable that, Charles the First and the foreign ambassadors were there to view that fortress with all its treasures and curiosities. . . .

Charles the Second was favoured with the exhibition of another scheme of this sort, by John Evelyn, a Fellow of the Royal Society at the time, and therefore not likely to participate in any matter which the scientific world of his day repudiated. But learned men of his time rather approved of all wonder-working automata than otherwise. Evelyn says in his Diary, under the date of 14th July,† 1668, that during an

---

* In this context, it's perhaps worth noting the full title of Worcester's book: *A Century of the Names and Scantlings of Such Inventions as at Present I Can Call to Mind to Have Tried and Perfected which (My Former Notes Being Lost) I Have, at the Instance of a Powerful Friend, Endeavoured Now, in the Year 1655, to Set These Down in Such a Way, as May Sufficiently Instruct Me to Put Any of Them to Practice.* It's possible, as Dircks says, that Worcester might in the intervening period have forgotten the details of his construction. On the other hand, it surely beggars belief that Worcester might have forgotten whether or not he'd created a machine capable of perpetual motion.

† In fact the relevant entry is dated August 14, not July 14 – an atypical slip from Dircks.

interview with the King: "I showed his Majesty the perpetual motion [device] sent to me by Dr Stokes from Cologne."

In 1690, in an appendix to an essay called *On the Mechanics of Effervescence and Fermentation*, **Jean Bernoulli** (1667–1740), the younger of the two great Swiss mathematician brothers, made a strong plea in favour of the possibility of *perpetuum mobile*, even while admitting that so far no one had been able to achieve it:

> [Y]et many [natural] philosophers reject the idea, unanimously asserting that Perpetual Motion cannot be communicated and cannot be invented; which opinion is nevertheless not of any weight, seeing that they rashly judge that no one should be listened to who boasts of having found out such a thing . . . I do not hesitate to assert not only that Perpetual Motion may be discovered, but that it has now actually been discovered, as will be confessed by anyone who reads these lines . . . Does not Nature herself (who is never said not to operate by mechanical laws) indicate Perpetual Motion to be possible? To recall but one instance, what is the constant flux and reflux of the rivers and seas but Perpetual Motion? Does it not all belong to Mechanics? Therefore, you must confess that it does not exceed the limits of mechanical laws, and is not impossible; what then hinders that following Nature in this, we should be able perfectly to imitate her?

It's really rather embarrassing to read this great scientist (as rendered by Henry Dircks in his *Perpetuum Mobile*, 1870) using the same arguments that crank theorists and inventors have used so many times over the centuries, rooted confidently in a sort of arrogant ignorance and a dismissal of the proofs of them damn' theorists – what the heck do they think *they* know, anyway?

Bernoulli's own device – which of course he never attempted to construct, leaving that as an exercise for the reader – consists essentially of a tall tube (open at both ends), a broad beaker, two fluids of different densities that mix readily, and a filter that permits passage of the lighter but not the heavier fluid. Mix the fluids, then put them in the beaker. Place the filter over one of the tube's open ends and place the tube vertically into the beaker, filter-end down. If you have calculated the relative heights and diameters of tube and beaker correctly in relation to the densities of the two fluids, you should find, Bernoulli reasoned, that the weight of fluid in the beaker pushes the lighter fluid up the tube and out of the top end. The liquid falls back into the reservoir, mixing there – the net effect being to create a perpetual flow.

Or not, as the case might be.

A claimed perpetual motion machine that actually worked – although it was not, alas, a perpetual motion machine, any more than a self-winding watch is a perpetual motion machine – was Cox's Timepiece, invented in the 1760s by the English inventor **James Cox** (c1723–1800), built by his compatriot automaton-maker John Joseph Merlin (1735–1803), and currently in the Victoria & Albert Museum. This relied on fluctuations in barometric pressure, as registered by a mercury column, to keep the mechanism wound. Even earlier, the Dutchman **Cornelius Drebbel** (1572–1633) had built a number of clocks on similar principles;* and the same notion underlies the modern Atmos clock, which relies on ambient temperature fluctuations.

Among the best known of all purported perpetual motion machines must be the series of devices created by **Johann Bessler** (1680–1745), also known as **Orffyreus**. The Landgrave of Hesse–Cassel and his cronies became convinced Bessler had discovered a perpetual energy source, and in 1716 gave him a well paid sinecure so he could work on the construction of his largest device yet. Some 3.7m in diameter and 45cm thick, this mighty wheel, once started, revolved at a steady speed of 26rpm for a period of weeks without further encouragement; in its official test, conducted by the Landgrave and various scientific dignitaries, including Willem 'sGravesande (1688–1742), it ran for 54 days unattended and was still going strong. The great Austrian architect Johann Bernhard Fischer von Erlach (1656–1723) was another witness, and testified:

> The wheel turns with astonishing rapidity. Having tied a cord to the axle, to turn an Archimedean screw to raise water, the

---

* Drebbel also (1620) built the first navigable submarine, an oar-powered vessel constructed at the behest of James VI & I.

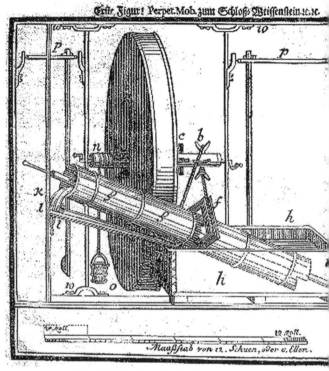

Bessler's (Orffyreus's) wheel

wheel then made twenty turns a minute.* This I noted several
times by my watch, and I always found the same regularity. An
attempt to stop it suddenly would raise a man from the
ground. Having stopped it in this manner [so that] it remained
stationary (and here is the greatest proof of perpetual motion),
I commenced the movements very gently to see if it would of
itself regain its former rapidity, which I doubted; but to my
great astonishment I observed the rapidity of the wheel
augmented little by little until it made two turns, and then it
regained its former speed. This experiment . . . convinces me

---

* This differs from other accounts, which say 25–6 or precisely 26.
Probably von Erlach just misremembered the number.

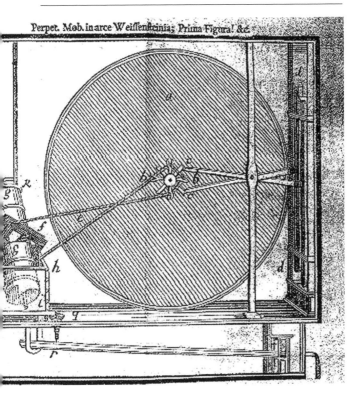

Perpet. Mob. in arce Weiſſenſteinſa; Prima Figura! &c.

more than if I had only seen the wheel moving a whole year, which would not have persuaded me that it was perpetual motion, because it might have diminished little by little until it had ceased altogether; but to gain speed instead of losing it, and to increase that speed to a certain degree in spite of the resistance of the air and the friction of the axles, I do not see how anyone can doubt the truth of this action.

Von Erlach was right in that his test was a better one than the (as it were) endurance trial to which the device was being put. As noted, it turned for 54 days unattended – the key word being "unattended". Supposedly in order to

prevent any tricksterish interference with its running, it was
kept behind locked doors; when the experimenters were
finally let back into the room they found the wheel still
moving. It seems not to have occurred to anyone that there
was no guarantee it had remained in motion during the
intervening period. It could have been clandestinely started
up mere minutes before the doors were opened.

What was the secret of the wheel's turning? Bessler
promised to reveal it to whoever paid him an inordinately
large sum of money; when all concerned shied away from
this, showing themselves wiser than many before and since,
Bessler refused to cooperate any further. On the Landgrave
suggesting a Dutch engineer be brought in to examine the
device, Bessler flew into a fury and smashed it. He lived
nearly three decades longer – eventually falling to his death
while working on another grandiose project, a vast windmill
commissioned by Karl I of Prussia – yet still he kept the
mechanism of his mighty wheel to himself. Other inventions came from him, but nothing that shattered the laws of
physics.

Bessler's taciturnity was not matched by that of one of
the Bessler family maids, Anne Rosine Mauersbergerin, who
in 1727 declared to the authorities that the mechanism
responsible for all of Bessler's wheels was a little less exotic
than claimed. A relatively small crank in the next room was
linked to the big wheel – whose support posts were not solid
but hollow – through a system of belts and gears; turning it
had been the shared task of herself, Bessler's brother,
Bessler's wife and Bessler himself. Naturally enough, the
scientists of the day – including 'sGravesande – refused to
believe a word she said, and Bessler's modern devotees do
likewise. As 'sGravesande summarized: "I know perfectly
well Orffyreus is mad, but I have no reason to think him an
impostor. . . . [I]f the servant says the above, she tells a
great falsehood . . ." In later centuries, whole hosts of
hoodwinked scientists were to react very similarly to the
feats of bogus spirit mediums, fraudulent inventors and
charlatan psychics: *As a distinguished scientist I would certainly
have noticed had any fraud been perpetrated . . .*

A design dreamed up in 1790 by one **Conradus**

Schiviers produced a crop of variations over the subsequent decades. A vertically mounted wheel has compartments into which heavy metal balls can fit. With all the compartments on one side of the vertical filled and all those on the other side empty, the wheel will naturally turn in the appropriate direction until all the filled compartments are at the bottom and all the empty ones at the top. In Schiviers's arrangement, however, this equilibrium is never permitted to occur: as each ball reaches the nadir it drops out through a hole in its compartment to land on an endless belt, which carries it to the top, where once more it falls into a vacant compartment . . . About a century later, an unknown inventor refined the idea such that the endless belt could be dispensed with: instead the balls were strung together like pearls on a necklace. So confident was this unsung genius that he included a brake in his designs in case the wheel spun out of control.

In 1812 a character called **Charles Redheffer** turned up in Philadelphia, seeking money from the city fathers for his device, which would run forever without fuel. One of the officials sent to investigate the machine became suspicious when Redheffer refused to allow anyone to come close, and realized the device was being driven by some source outside the room. Rather than reveal the hoax, this man invited Redheffer to City Hall; by the time Redheffer got there, a working duplicate of his machine had been set up. Redheffer abruptly decamped to New York City, where it was his misfortune that his machine attracted the attention of the renowned engineer Robert Fulton (1765–1815). Fulton, like the Philadelphia official, soon spotted where the motive force was coming from, and led a band of interested citizens upstairs to where a man was boredly turning a crank from which a cord led down through the floor. The citizens proceeded to "dismantle" Redheffer's machine.

Sir William Congreve (1772–1829), of Congreve's Rocket fame, tried his hand at a very individualistic perpetual motion device, and in 1827 got a patent for it. There's no evidence he ever tried to build a prototype, which fact undermines his frequent claims the device would work. It depended on the phenomenon of capillary lift, whereby

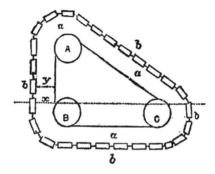

Congreve's inclined plane, with an endless belt of
sponges being immersed in water

surface tension drives water upward, against the force of
gravity, through very fine fissures or tubes; it's largely
because of capillary lift that water from the soil can reach
the tops of trees. In Congreve's device an endless belt of
sponges is placed over an inclined-plane setup whose lower
part is submerged in water. Forming a further layer outside
the chain of sponges is a chain of metal weights. When the
sponges are on the inclined plane, the weights squeeze out
of them the water they have acquired during their period of
submersion; they're thus relatively light. Those sponges that
have reached the top of the inclined plane descend vertically back into the water. They become saturated even
before they reach the water, however, because of capillary
lift, and of course the metal weights are no longer pressing
down on them to squeeze the water out. Because they are so
much heavier than their counterparts on the plane, they
drag those counterparts up that slope. And so *ad infinitum*.

An inventor called **Harry Prince** gained a UK patent in
1866 for a submerged wheel bearing on its outer rim a
succession of air bags and weights arranged such that the
weights squashed the bags, evacuating them, at one point in
the rotation; this was countered by the upward thrust elsewhere on the wheel of the undeflated bags, which were
buoyant in the water. The turning wheel operated the pump

that reinflated the emptied bags. Henry Dircks commented on this item: "This singular specimen of modern mechanical stupidity was duly honoured with a final and complete [patent] specification."

Just a few years later, on the other side of the Atlantic, the Chicago inventor **Horace Wickham** gained a US patent in 1870 for a rather interesting device, a sort of perpetual motion seesaw whose rocking could be converted into useful power via a series of gears. A hollow beam is pivoted at its centre. A heavy ball runs along the inside of the beam. As the ball passes the pivot its weight rocks that half of the beam downwards. On reaching the beam's end, the ball drops through a hole and into a set of tubes along which, neatly ignoring fundamental principles, the ball runs uphill back to its starting position, ready for another run.

A far more interesting perpetual motion device than most of its contemporaries was that developed in the 1880s by the UK-born Washington DC inventor **John Gamgee** (1831–1894).* What's going on in a steam engine is, essentially, that heat energy from the furnace is being converted into kinetic energy. Many people had observed that the steam engine bore all the potential to be a perpetual motion machine aside from the fact that it was hopelessly wasteful: only a fraction of the heat energy was converted into kinetic energy, the rest being lost through various inefficiencies. Gamgee's great brainwave was to construct the equivalent of a steam engine but using ammonia instead of water. Water has to be heated into its gaseous form (steam); ammonia, by contrast, is a gas at room temperature (its boiling point is –33.34°C). Gamgee's device was called the zeromotor because designed to operate at 0°C. If you introduced liquid ammonia into the device, heat from the environment would vaporize it, enabling it to drive a piston. In the other cycle, ammonia vapour expelled into a reservoir would condense

---

* Some accounts describe Gamgee as a British university professor who happened to work in Washington DC. This seems improbable. The ODNB gives him as a "veterinary surgeon and inventor". He likely buffed up his *curriculum vitae* in hopes of impressing US acquaintances.

and, as a liquid, run back into the boiler. What Gamgee didn't realize was that the amount of energy required to cool the ammonia gas down enough for it to become a liquid was far greater than any energy you could extract from the movement of the piston.

While that might seem obvious to anyone who's calculated how much the fridge and freezer contribute to the electricity bill, it wasn't obvious to Benjamin F. Isherwood (1822–1915), Engineer-in-Chief of the US Navy, who in 1881 had no hesitation in recommending to the Secretary of the Navy that money be put into the further development of the device:

> The possibility of the invention of a new motor of incalculable utility would seem to be established . . . Prof. Gamgee's invention . . . is not that of a machine for the application of power, but for the immensely more important purpose of generating power itself, so that, strictly speaking, it includes as a basis all other machines.

Isherwood claimed to have examined one of Gamgee's prototypes to ensure that "any part of the ammonia which entered the cylinder as a gas left it as a liquid" – the point noted above – and declared himself satisfied this was the case . . . but with a qualification: "so far as the form of the apparatus allowed any observation to be made". It seems Gamgee was no fool.

Much later, in 1978–9, a very similar principle was at the heart of the pump motor into which Stewart Energy Systems of Idaho was proposing US farmers sink their money; the machine invented – or, to be accurate, never *quite* invented – by **Robert C. Stewart** was much like Gamgee's except using freon rather than ammonia, and the intention was that it could be adapted for irrigation purposes, so farmers could save enormously on their energy costs. The minor technological detail that Stewart was never quite able to master before the Security and Exchange Commission shut Stewart Energy Systems down (by which time the company had pulled in some $3 million) was the heat-exchange engine required to extract energy from water by chilling it. This is a technological detail that has confounded better inventors than Stewart. Needless to say, he regarded himself as a martyr, and as recently as late 2008 was apparently declaring his faith in his device – for which, it was claimed, he now had corporate backing. It'll be interesting to see what comes of this.

Reverting to the late 19th century and the first decades of the 20th, by the 1890s the great inventor **Nikola Tesla** (1856–1943), who during the 1880s had devised the alternating-current system of electricity supply we use in our homes today, was investigating the possibilities of a new type of electrical generator that would run without fuel. Of course, so much myth surrounds the figure of Tesla that it's very hard to separate fact from fiction – a situation he himself did much to encourage – but it does seem certainly the case that he spent several decades working on this problem. He also said he'd actually invented such a generator: in a 1902 letter to Robert Underwood Johnson (1853–1937), editor of the *Century Magazine*, Tesla wrote – in response to the news that a Canary Islands engineer called Clemente Figueras claimed to have invented a fuel-less electricity generator – that he himself had already developed such a device.

And much later, in 1931, Tesla reported in an article for the *Brooklyn Eagle*:

> I have harnessed the cosmic rays and caused them to operate a motive device. . . . More than 25 years ago I began my efforts to harness the cosmic rays, and I can now state that I have succeeded.

A couple of years afterwards, in a 1933 article for *New York American*, Tesla stated:

> This new power for the driving of the world's machinery will be derived from the energy which operates the universe, the cosmic energy, whose central source for the earth is the sun and which is everywhere present in unlimited quantities.

Of course, this could be interpreted – and undoubtedly has been a million times over by the Tesla fraternity – as a reference to nuclear fusion. It is difficult, now, to establish what it was he was after; even so, Oliver Nichelson made a good attempt in a 1991 paper called "Nikola Tesla's Later Energy Generation Designs".* What's evident in all the designs Nichelson considers, and in all the relevant Tesla statements he cites, is that, while the headlines might have talked about fuelless engines, the story was really one of discovering means of tapping energy already naturally present, such as that of the sun. There is, obviously, no violation of physical laws involved here.

Other inventors have, of course, been not so scrupulous. "Call unto me, and I will answer thee, and shew thee great and mighty things, which thou knowest not." So reads *Jeremiah* 33:3, and this verse gave its name to the device invented by **Arnold Burke** of Temple, Texas, and into which he persuaded the usual hordes to put their money. His "self-contained hydroelectric power system" relied for its functioning on a "self-acting pump" that he claimed to have invented earlier in order to drain a goldmine. Initially he got funding from one of the big dairy co-ops, but they backed out when he was unable to demonstrate a working model. Next he sold distributorships to private individuals, targeting fundamentalist Christians in particular – hence the naming of the device as Jeremiah 33:3. In 1979 the state of Texas stepped in to stop all this, and at trial was able to insist the gadget should be examined by an independent expert. Oddly, during the inspection it wasn't the expert, engineer David Kehl, but the Assistant Attorney General, Roy Smithers, who spotted a furtively placed electrical wire. Sure enough, the design of Burke's "working model" incorporated a hidden electric pump.

The resulting fraud trial resulted in an 11:1 hung jury – a narrow escape for Burke. His claim was that he'd installed the electric pump before the inspection in case Kehl and Smithers, on discovering the principle of the machine's operation, stole the idea. What's bizarre is that, between the discovery of the deception at the end of 1979 and the fraud trial in Spring 1980, Burke's pious support-

---

* Nichelson has elsewhere speculated that the 1908 Tunguska Event, in which a large acreage of Siberia was flattened, might have been an accidental by-product of one of Tesla's telecommunications experiments. We should not hold this against him.

ers were able to raise $250,000 for his defence from investors old and new!

Eventually the patent offices on both sides of the Atlantic wearied of the task of poring over blueprints in order to demonstrate the flaws in perpetual motion devices, and clamped down. In 1911, for example, the US Patent Office* issued a memo that began:

> The views of the Patent Office are in accord with the scientists who have investigated this subject and are to the effect that such devices are physical impossibilities. The position of the Office can be rebutted only by the exhibition of a working model.

This dramatically cut down the number of perpetual motion patent applications. Some inventors, however, turned their ingenuity to naming and describing their devices such that the examining officers might not realize exactly what was being patented. Others, whose motives were more genuinely fraudulent, were able to devise models that did actually hum and click and whirr, but not for the reasons stated in the patent application.

There were also designs whose applications for a patent without a working model were, inventors might quite reasonably argue, legitimate because they depended on technologies that had not yet been realized but which might very well one day be feasible. Even though the inventor didn't claim to have developed the technology itself, the *application* of it was an invention in its own right. For example, consider a machine that depended on the discovery of a material which blocked the pull of gravity – like the "cavorite" H.G. Wells imagined for his novel *First Men in the Moon* (1901). Place a vertical wheel such that one half is above open ground and the other above a sheet of cavorite,

---

* Long before this the French Royal Academy of Sciences had in 1775 decreed it would "no longer accept or deal with proposals concerning perpetual motion".

give the wheel an initial shove, and – *voilà!* – assuming the pivot isn't too sticky, your wheel should spin forever. Gravity's pulling down on every element of one half of the wheel, but there's no gravity acting on the elements of the other half to deter them from rising.*

The same stricture could be applied to any of the notions today attached to the excitingly futuristic field of zero-point energy. Even the hardest vacuum is not in fact empty: it's filled with "virtual particles", which are particles that just happen not to exist at the moment. In the mysterious universe revealed at the quantum level, however, it's possible for these virtual particles to spring into existence spontaneously without any conservation laws being violated so long as they do so in pairs, comprising a particle and its antiparticle; that way (to simplify hugely) all the particle's plus signs are cancelled out by all the antiparticle's minus signs, so the net result is zero. In the ordinary way, what happens in this process of pair creation is that the two annihilate each other immediately, but it's just feasible the annihilation might not happen – or might, using some supertechnology, be stopped from happening. In this latter event you'd have a source of free energy, courtesy the universe. The primary exponent of "vacuum engineering" or "vacuum farming", supposedly the up-and-coming solution to all our energy problems, is Harold Puthoff (b1936), well known for his efforts with colleague Russell Targ (b1934) of the Stanford Research Institute to persuade business and the military of the exploitable potential of psychic powers. Nobel physics laureate Steven Weinberg (b1933) did the math and came up with a result that might have been expected to dampen Puthoff's enthusiasm: if you mined a volume of vacuum the size of the earth you would, if lucky, be able to extract the same amount of usable energy as there is in a gallon of gasoline.

---

* Responsible for this thought-invention was George Rideout Sr of the Babson Foundation. We'll encounter the Babson Foundation shortly when we look at antigravity research: see page 223.

An 1889 photo showing a self-satisfied John E.W. Keely in his lab

If perpetual motion itself is accepted to be an unrealizable dream, what about the next best thing? – devices that require almost zero fuel, or a fuel that's in plentiful, cheap supply.

In the latter part of the 19th century **John E.W. Keely** (1827–1898) extracted large sums of money from investors for his "vibratory generator with a hydro-pneumatic pulsating engine", which performed numerous remarkable feats – twisting metal bars, or running unattended for weeks on end – and all this apparently fuelled by just small quantities of water. In order to develop his engine, which he called the Liberator, he launched the Keely Motor Co. in 1872 with a capitalization of $5,000,000. By 1886 his stockholders had grown dissatisfied with his failure to produce a version that could be put to practical use, and they dissolved the company.

Keely's cause was taken up by the philanthropist Clara
Jessup Bloomfield Moore (1824–1899), a rich widow whose
worthier causes included the Temperance Home for
Children. She put her money where her enthusiasm was,
and so he was able to continue living the good life. Perhaps
she became suspicious, or perhaps her friends were sceptical, or perhaps she just wanted to show off her pet inventor,
but in 1895 she asked Addison B. Burk (1847–1912),
President of the Spring Garden Institute (a Philadelphia
technical college), to investigate the device. He reported he
believed the machine a fake and Keely a charlatan. Mrs
Moore decided to ignore this and continued to invest in
Keely. At least for another year: in 1896 she called in the
UK pharmaceutical chemist Wentworth Lascelles-Scott.*
Astonishingly, Lascelles-Scott was persuaded to issue a clean
bill of health: "Keely has demonstrated to me, in a way
which is absolutely unquestionable, the existence of a force
hitherto unknown." This judgement was vehemently
contested by the electrical engineer E. Alexander Scott,
who'd visited Keely's premises on more than one occasion
and was profoundly sceptical of the inventor's claims. Mrs
Moore suggested the two scientists conduct a joint investigation; when Keely refused to cooperate with one of Scott's
requests, it seems Lascelles-Scott reversed his earlier opinion. Even Mrs Moore's credulity was jolted; she reduced
Keely's allowance to a mere $250 a month – the equivalent
of about $7000 today.

Only after Keely's death did investigators find in his
basement the hydraulics and compressed-air equipment
really responsible for driving the wondrous machines.

It's surprising Keely bothered with a power-source as
mundane as compressed air, for he is reported to have
discovered a "sidereal" force known to the inhabitants of
Atlantis as "Mash-Mak". Mash-Mak, if correctly exploited,
could provide an energy source beside which even our
modern nuclear energy would seem no more potent than an
elastic band. However, humanity was not ready for this
major breakthrough and so, we are told, Keely was "myste-

---

* According to the *Wellesley Index to Victorian Periodicals*, Lascelles-Scott
had already, in 1894, published on the subject of Keely's inventions.

riously silenced". According to his own claims, his publicly demonstrated devices used the luminiferous aether's vibratory energy. The "vibratory" part of this did explain why sometimes starting his devices required him to play a snatch of music on a harmonica or other instrument that might be to hand. A typical pronouncement of Keely's, belligerent and opaque at the same time, reads:

> Science, even in its highest progressive conditions, cannot assert anything definite. The many mistakes that men of science have made in the past prove the fallacy of asserting. By doing so they bastardize true philosophy and, as it were, place the wisdom of God at variance; as in the assertion that latent power does not exist in corpuscular aggregations of matter, in all its different forms, visible or invisible.[*]

Although Keely was posthumously exposed as a fraud, people continue to emulate Clara Moore. *The Journal of Sympathetic Vibratory Physics*, dedicated to the furtherance of Keelian physics, was issued regularly between 1985 and 1990, and you can still purchase a complete run of it from those friendly folk at Delta Spectrum Research.

Joshua Gulick, in a 1999 article called "Gravitation and Distortion Systems", appears to be making the claim that what Keely had stumbled onto were the basics of superstring theory.

Louis Enricht (*c*1846–1924) was just another small-time conman until 1916, when he was aged 70.[†] The reason for his shooting to fame was his claim to have invented a chemical compound that could be added to ordinary water to make a cheap gasoline substitute. Gas stations faced bankruptcy: who'd need a gas station when you could stick enough of Enricht's compound in your pocket to last a month?

---

[*] Opening of Keely's contributed chapter "Latent Forces in Interstitial Spaces, Electro-Magnetic Radiation, Molecular Dissociation" in *Keely & His Discoveries: Aerial Navigation* (1893) by Mrs Bloomfield Moore.

[†] One of the more scientific of his earlier scams, which he operated profitably in 1912, was to persuade New York City investors he'd discovered a way to make a concrete-like building material out of mud, sand, ashes or even sawdust.

On April 11 1916 – ten days late, one might think –
Enricht called a press conference at his home in
Farmingdale, Long Island, to announce his discovery. The
journalists came for a kook story to fill the bottom of a page
somewhere, and initially mocked him. Then, before their
very eyes, Enricht mixed water and a little of a green liquid,
poured the mixture into the tank of his car, and drove
around Farmingdale. A few days later William Haskell,
publisher of the *Boston Herald*, arrived in town, and Enricht
repeated the demonstration. With the *Herald*'s assistance,
Enricht became a national figure. Before long Henry Ford
(1863–1947) was hooked. While Ford was still negotiating –
but after he'd put down an "earnest" of $1000 – the Maxim
Munitions Corporation announced it had bought the exclu-
sive rights to manufacture Enricht's compound for a
reported million dollars plus 100,000 Maxim shares – which
shares were of course set to go through the roof once
manufacture began. Initially Ford created a furore over
Enricht's double-dealing, but then he suddenly backed
off, presumably on scientific advice. Not long afterwards,
the company's stock having already doubled in value,
Maxim abruptly denied its earlier announcement – indeed,
not just denied it but tried to pretend it had never been
made.

In November 1917, some months after the US entered
WWI, the railroad financier Benjamin Yoakum (d1929)
stunned the nation by announcing that Maxim's renuncia-
tion had in fact been true. Enricht had been *triple*-dealing,
and in April of that year had partnered himself with
Yoakum, who was paying all the bills, to found the National
Motor Power Company. Yoakum was now going public with
the arrangement because Enricht was, Yoakum claimed,
refusing to give the formula to the US Government, as was
surely the duty of any US citizen in wartime, and had even
been in contact with the German Government with the idea
of selling the formula to *them*. Enricht was able to lie and
finagle his way out of criminal charges, and anyway the US
Government showed little interest in his formula.

Enricht perpetrated further technological scams. Other
"inventions" were a way of extracting nitrogen from the air

for use in explosives and pesticides and a method of deriving gasoline from peat. It was the Enricht Peat Gasoline Corporation, which he founded in early 1921, that led to his downfall: the company attracted plenty of investors but, when no peat-derived gasoline was forthcoming, they sued. Tried for fraud, Enricht was sent to Sing Sing. In 1924 his sentence was commuted to "time served"; he died soon after his release.

The Armenian-born Bostonian **Garabed T.K. Giragossian** modestly dubbed his marvellous device the Garabed. So certain was he that this machine was indeed a source of free energy that he bypassed the US Patent Office and in 1917 went straight to Congress. He explained to the House Committee on Patents that inventors of the past, on making similar breakthroughs, had been sinisterly silenced or sidelined; he begged protection. Accordingly, early the following year the House passed, and President Woodrow Wilson signed, Public Resolution #32, which gave Giragossian the protection he'd sought – on condition that a scientific committee vouched for the worth of his machine.

A few months later four MIT scientists vetted the Garabed. They delivered their report the following day, and apparently it ran to just one paragraph. The much-vaunted Garabed was a flywheel, which was powered up to speed by human effort and then kept spinning by a little electric motor. If a braking dynamometer were used to bring the flywheel to an abrupt halt, the dynamometer registered a power many times that of the electrical motor. Giragossian had been too ignorant of basic mechanics to know that power is a function of energy *and time*: the high reading on the dynamometer was a result simply of the speed with which the wheel had been brought to a halt; if he'd slowed it down more gently the energy reading would have been much lower, though recorded over a longer period of time.

It's difficult to underestimate the levels of scientific illiteracy displayed by many politicians. Giragossian of course refused to accept the MIT scientists' verdict – they just *didn't understand the principles*, you see – but less forgivable is the fact that, while Congress never did provide funding for the

device, it wasted taxpayer money on further hearings about
it in 1923 and 1924, and was still voting on relevant resolu-
tions as late as 1930.

There was another aspect to Giragossian's work. The
August 1926 issue of *The Herald of Christ's Kingdom* reported
completely uncritically on this magnificent new technological breakthrough, prefacing its report thus:

> The fulfilment of the prophetic pictures, symbols, and visions
> of the Bible, descriptive of the coming Age of man's return to
> Paradise, must without doubt involve discoveries, inventions,
> and revolutionary changes in the affairs of mankind,
> compared with which, those of the present will seem small and
> insignificant. The providing of food, raiment, and shelter for
> all the awakened dead, their uplift out of every kind of weak-
> ness, degradation, and degeneracy, their final establishment in
> the perfect state of the image and likeness of God in a restored
> and transformed earth, from which every vestige of the curse,
> thorn, thistle, and destructive blight and insect, etc., will have
> been removed, will necessitate the employment of forces and
> agencies at present entirely unheard of.

Yes, well, but isn't that the sort of mopping up you'd expect
to be done by the Good Lord Himself as part and parcel of
the Second Coming? Apparently not:

> There comes to our attention at this time an article of recent
> publication describing a new and marvelous invention that it
> is alleged is capable of what would seem to be impossible of
> achievement. Knowing that many of the readers of this journal
> are observing with special interest the signs of our day that
> mark the imminence of the new dispensation and the
> Kingdom of God, we quote liberally from the article in ques-
> tion . . .

As indeed they do, regurgitating what reads like a press
release produced by Giragossian himself. This must be the
ultimate accolade for a bogus machine: to be recommended
for use in the humanitarian relief operation after
Armageddon.

Another wondrous device supposedly capable of giving
out more energy than required to power it was the Kenyon

Alternator, demonstrated by Solar World Inc. of California in a celebrity-draped public launch on March 22 1979. In this device, the brainchild of US physician **Keith Kenyon**, permanent magnets were mounted on the rim of a wheel that spun inside an array of coils, the purpose being to derive from the coils more electricity than required to spin the wheel. The demonstration certainly seemed convincing enough, with a bank of a hundred lightbulbs glowing brightly to impress the spectators while on a screen were displayed the figures for the current input and output: apparently a measured input of 5936W from a five-horse-power motor was yielding a measured average output of 7512W – a gain of about 26.5%.

What seems not to have been evident to the officers of Solar World was that the simple meters they were using were designed to work with the sinusoidal waveforms typical of an AC electrical supply. Unfortunately, the output of the Kenyon Alternator wasn't an AC electrical supply, and so the readings were meaningless.

When it was suggested the best way of finding out whether there was any real net gain in energy would be to hook up a Kenyon Alternator such that it powered the motor that in turn drove the Kenyon Alternator, there were declarations that exactly such a project was on the blocks. There, however, it seems to have remained.

The Newman Energy Machine, devised by **Joseph Westley Newman** of Lucedale, Mississippi, was similarly an electrical device that claimed to have higher output than input. It was first brought to major public attention in 1984 when Dan Rather featured it on CBS News. When Newman had tried to patent his battery-operated gadget in 1979, however, the US Patent Office rejected it, claiming it was just a perpetual motion machine in disguise. (As with Kenyon's device, if the device produces more energy than is required to run it, then one should be able to set it up to power itself and stay running forever.)

Newman denied this and sued, and eventually, in 1984, the case came to court. The judge appointed the electrical engineer William E. Schuyler Jr to investigate the device, and surprisingly Schuyler gave a positive report:

Evidence before the Patent and Trademark Office and the Court is overwhelming that Newman built and tested a prototype of his invention in which the output energy exceeds the external input energy; there is no contradictory factual evidence.

Even so, the Patent Office stuck to its guns and declined to issue a patent. Eventually the contretemps returned to court in 1986.

By then the device had been, at the Patent Office's request, thoroughly tested by the National Bureau of Standards (NBS),* whose scientists gave it a thumbs-down; according to their measurements, far from stepping up the energy, the motor was, under varying settings, registering efficiency levels of only 33%–73%. In court Newman's lawyers offered as expert witnesses two engineers (Milton Everett of the Mississippi Department of Energy and Transportation, and Ralph M. Hartwell II, a TV station engineer) and a physicist (Roger Hastings of Unisys), all of whom said in essence that the NBS scientists were incompetent. In addition, both engineers said they'd carried out informal but positive tests on Newman's prototypes; Hastings had done the same but, more interestingly, explained something of the theory behind Newman's device.

Newman had become convinced there must be a mechanical explanation for electromagnetic fields, and eventually theorized that they consisted of gyroscopic (i.e., spinning) particles travelling at light speed along the flux lines. He was then able to build up an entire theory of electromagnetism based on the interactions between these gyroscopic "massergies". Since electromagnetic fields involved particles in motion, there must be a way of tapping this kinetic energy. Further, the massergies did not come from nowhere: they must have been derived from the substance (e.g., the wire coil) producing the electromagnetic field. The mass lost by that substance, though tiny, should be manifested as a very great deal of energy, according to the famous $E = mc^2$ relationship.

---

* Ancestor of today's National Institute of Standards and Technology, NIST.

An electric current, in Newman's view, is a flow of gyroscopic massergies moving at light-speed along a wire, aligning electrons as they go. If the wire is sufficiently long, and if the current is an alternating one, the massergies, even travelling at light-speed, will not have time to reach the end of it before the direction of their travel is reversed; accordingly, they'll become trapped within the wire, even as further massergies are pumped into it from the power supply. This should permit a build-up of electromagnetic energy in the wire. As it happens, the coils in the Newman motor's alternators contained considerable lengths of wire – up to 55 miles (88.44km) – and the switching was very rapid. This prompts an immediate question: Was the switching *really* so rapid as once every 2.8×10^{-4} seconds, which is the time light takes to travel 88km? It's hard to credit.

Newman himself, cited by Robert Schadewald in his essay "The Perpetual Quest" (see Bibliography), has offered a rather folksier explanation while also pressing his case:

> I have persistently stated that this 100% conversion process was achieved by tapping into the basic building block of matter – the energy (gyroscopic particles) comprising a magnetic (electromagnetic) field. I have disclosed that this energy in a magnetic (electromagnetic) field was literally "a river of energy" that could be tapped for immediate benefit of humanity without pollution to the environment or the human race, would end hunger, would make space travel commonplace, and much more. Also this technology will do more to bring about "World Peace" than all the Kings, Queens, and politicians that have ever lived.

Whatever: The NBS refused to back down, claiming its experts knew what they were doing, and eventually, in 1988 – such was the extent to which the wranglings had dragged on – the judge came down on the side of the Patent Office. Newman had not assisted his case by issuing a press release in 1987 describing the judge as, in its milder moments, "a criminal of the worst kind" and demanding his impeachment. Newman also announced his intention to run for US President as the candidate of the Truth and Action Party.

Newman's case has been advanced a number of times in Congress, but each time the representatives have declined the opportunity to force the Patent Office to recognize his invention. And he's still active today: his website is full of exhortations to buy relevant DVDs (or watch shorter presentations on Google Video) plus plentiful references to God.

But what of all the respectable scientists who, like Hastings, have testified that the operation of Newman's motor is above board? We have to recall that marvellous machines have throughout their history attracted positive testimonials from highly esteemed scientists – remember 'sGravesande (page 198), Bernoulli (page 194) and others. From time to time even the spoonbending feats of Uri Geller (b1946) and similar "psychics" have been described as genuine by scientists, while numerous distinguished scientists were misled during the glory days of Spiritualism, in the late 19th and early 20th centuries, among them Sir William Crookes (1832–1919) and Sir Oliver Lodge (1851–1940).

One very noteworthy scientist to be caught up in the whole energy-out-of-nowhere business was **Eric Laithwaite** (1921–1997), Professor of Electrical Engineering at Imperial College, London, and designer of the Maglev and of the world's first high-speed train. In 1973 Laithwaite demonstrated at a lecture to the Royal Institution a supposed antigravity device he'd built. It weighed about 25kg and looked like a big gyroscope mounted on the end of a pole. When the gyroscope was at rest he could barely lift the device, but when he spun the gyroscope up to speed he was able to lift the assembly over his head one-handed. As he did so he foolishly made a remark to the effect that he was violating Newton's Laws of Motion. Later it was to be suggested that it was this remark rather than any analysis of the physics involved that led the Royal Institution to refuse to publish his lecture, the first time this body had taken such a step.

Gyroscopes *do* seem to act in violation of the laws of physics – not least gravity! Since angular momentum, which is a directional property, is maintained in a rotating object, if you hold a spinning gyroscope in your hand and try to shift the direction of its axis you can feel the device's resistance to your action almost as if the gyroscope were alive. Similarly, if you set your spinning toy gyroscope atop one of those little plastic towers that normally come as part of the package, it will adopt positions that seem to defy gravity as its axis pivots. In fact, of course, all of these properties are in direct consequence of the Newtonian Laws.[*]

The physics of rotating gyroscopes fascinated Laithwaite, and he spent years investigating them mathematically. In the end he was able to give a watertight proof that indeed Newton's Laws were being obeyed, but this didn't stop him from believing that gyroscopic behaviour could somehow be harnessed to produce a reactionless drive – and towards the end of his life he was granted a US patent for exactly such a drive, which he took to prototype stage. Since one might say that Laithwaite's career was based on successfully inventing things that everybody said wouldn't work, there has inevitably been a sort of cultlike interest in this device ever since. It should be noted, though, that Laithwaite himself calculated it would require so much fuel that it offered little advantage over more conventional drives.

As recently as January 2009 I was astonished to discover on eBay the following item:

> ## PERPETUAL MOTION
> ## ANTI GRAVITY DEVICE
> no batteries required
> "the world's most efficient spinning device"
> brand new in gift box, length approx 13cm
> color may vary from the color in the photo

For a mere $AU20.00 (including shipping), could I resist this bargain? Well, yes, I found that I could.

---

[*] General Relativity suggests that a gyroscope spinning at relativistic speeds might affect local gravity. It's rather difficult to imagine something spinning that fast without immediately ripping itself to pieces. Besides, at relativistic speeds the gyroscope's mass would rise towards infinity.

Some people really believe the technology of antigravity is just around the corner, if not already in existence.

Predictably, NASA hears from many amateur inventors about the breakthroughs they've made that might, if properly exploited, lead to a new space drive, or at least to a cheaper way than rockets of getting stuff into orbit. These ideas are passed to NASA's Breakthrough Propulsion Physics Project, whose staff have wearily compiled lists of known fruitless lines of approach. Most of the proposals that arrive can be dismissed fairly rapidly by consulting these lists.

The three most common categories into which the proposals fall are gyroscopic antigravity, electrostatic antigravity, and oscillation thrust. Gyroscopic antigravity is the principle behind the researches of Eric Laithwaite (see above). In devices that rely on the notion of electrostatic antigravity, typically a high voltage is passed across a capacitor of special design; this causes the capacitor to lift off the surface. What seems to be happening is that ions passing from one capacitor electrode to the other may generate an air flow. If the electrodes are suitably positioned, this flow may produce an upward thrust. For more on electrogravitics see page 227.

The type example of an oscillation thruster is the notorious Dean Drive, devised in the late 1950s by mortgage appraiser Norman L. Dean – notorious because it captured the attention of the editor of the science fiction magazine *Analog*, John W. Campbell (1910–1971), who plugged it mercilessly to his readers as a great technological breakthrough. Many readers apparently built their own models and discovered that, yes, when the power was switched on the gadget juddered around the table or floor; even more excitingly, if placed on a bathroom scale it definitely appeared to lose weight.

The principle of the Dean Drive – and of oscillation thrusters in general – relies on impelling a weight or weights very fast in one direction and then allowing it or them to return slowly to the original position, where the cycle begins again. Imagine you've set the gadget on a tabletop, with the direction of the weights' motion being horizontal. The jolts caused by the weights' high-speed movement will jerk the device forward. The gentler return of the weights to their starting point won't have such an effect, because of the friction between the device and the tabletop. Over time, the net effect will be to take the machine from one side of the tabletop to the other. The Dean Drive might thus be adaptable to produce the most uncomfortable car ride you've ever had in your life, with an astonishingly high fuel consumption; but in space, where there'd be no friction between spaceship and its surroundings, you'd get just the discomfort.

With all of this in mind, it's disconcerting to discover a September 28 2000 SPACE.com report by Jack Lucentini that in the previous year NASA had given a $600,000 contract to Superconductive Components Inc. of Columbus, Ohio, to build a prototype antigravity machine. In answer to critics who said such a machine was a physical impossibility, Ron Koczor, Assistant Director for Science & Technology at the Space Science Laboratory, commented: "To say this is highly speculative is probably putting it mildly." On the other hand, if it would reduce the drag of gravity on launching spacecraft by even a little it would save such huge sums of money that $600,000 was a worthy gamble.

The device concerned was based on research published during the 1990s by the Russian engineer Yevgeny Podkletnov and colleagues. They had set a 30cm disc of a superconducting ceramic rotating very rapidly in a magnetic field; there was a definite gravity-reduction effect above the disc, they reported. In 1996 NASA put teams from the Marshall Space Flight Center and the University of Alabama at Huntsville to work together on the effect, but the research effort fell apart in acrimony for reasons that are disputed. In that same year Podkletnov was asked to vacate his position at the Tampere University of Technology, Finland, because of the controversial nature of his antigravity claims. In 1997 he declined an invitation from NASA to visit Huntsville for a consultation. Nonetheless, Superconductive Components's work on

NASA's 1999–2000 project was being done with his cooperation.

Unfortunately, the results were "inconclusive". So the following year NASA invested *another* $600,000 in trying to get the effect to work. Since then there's been a certain silence from NASA on the subject. Reports in the early 2000s that Boeing and BAE were independently researching the possibilities of an antigravity drive are, so far as I can establish, unconfirmed.

World peace will come only as the spirit of Jesus grows in the hearts of man and as the principles of birth control are taught to overcrowded nations and the latent power of gravity is used as freely as air, water and sunlight.
– Roger W. Babson, *Actions and Reactions*, 1950 (2nd rev edn)

In a sense it's probably all the fault of Albert Einstein that the field of antigravity is such a wide open territory for the unorthodox theorist. Before Newton, gravity seemed a fairly commonsensical sort of force: if you dropped something it fell to the ground because that was its nature, it being of the same sort of stuff as the earth and thus obviously eager to associate with it. The new idea of gravity that Newton brought in offered far more of a stretch to common sense, with its conception of a force operating over colossal distances between objects that seemed to have nothing to do with each other. Even so, people could eventually get their heads round the notion – could *visualize* it. But the version of gravity that Einstein proposed was a maze of mathematics, and thus went against the time-honoured dictum that all true science should be comprehensible to anyone. Even the popular analogy of objects behaving under gravity as if they were rolling around on a rubber sheet didn't seem to have much relevance to what was going on when someone stood on your foot. Hence there was a sudden upsurge of amateur cosmological theories which had in common the rejection of Relativity; the irony is that many of these are considerably harder to understand or visualize than the supposedly more complicated cosmologies emerging from orthodox science. And hence, too, the flourishing of the notion that, if Einstein was wrong because difficult to understand, then that must mean it should be pretty easy to find some way of reversing or negating the thing at the heart of his General Relativity: gravity.

Recent notions in orthodox cosmology and most particularly (as part of attempts to explain why the universe is apparently expanding faster than it "should be") the invocation of a force called dark energy that repels bodies *from* each other in the same way that gravity attracts them *to* each other, have obviously generated some excitement in antigravitic circles . . . especially since another term that has been used for dark energy is repulsive gravity. Yet, while very little is as yet known about the properties of dark energy (whose very existence is anyway still at the hypothetical stage, although the evidence is strong), what it does *not* seem to be is the kind of "negative gravity" the antigravitists pursue and theorize about.

To analogize: Simply because the weak nuclear force repels and the strong nuclear force attracts does not mean the two are opposites, inverses or complements of each other; in fact, the weak force seems to have more to do with the electromagnetic force than with the strong force. Similarly, just because gravity attracts and dark energy repels doesn't imply the two are mirrors. Gravity appears to be born of the geometry of spacetime; it's difficult to construct a model in which dark energy could *also* be born of the geometry of spacetime.

In "Chronology of US Antigravity Research" (*Antigravity News and Space Drive Technology*, Jan/Feb 1998) James E. Cox listed some highlights and breakthroughs. Here's a selection (all *sic*):

> **1896** - John Keely [see page 207] demonstrates a sympathetic vibratory flying machine to the US Army which flew 500MPH.
> **1943** – Philadelphia Experiment confirms Einstein's modified Unified Field Theory.

1947 – Roswell crash in New Mexico yields bonanza of extraterrestrial antigravity propulsion technology.

1950's – In general, there are a flurry of news accounts of amateur inventor's who claim success in building working flying saucer models.

1960's – Many American companies exhibit a euphoria of belief that control of gravity is eminent – Glenn Martin, etc.

1964 – Dr. Erwin Saxl publishes in NATURE anomalous weightless data on his charged and shielded torsion pendulum experiments.

1975 to 1985 – There is a stretch of time here where little public information on antigravity is available – probably a period of suppression or quiet research.

And so on.

The most substantial body of work done towards devising an antigravity machine came about as a result of the obsession of the US businessman and philanthropist Roger Babson (1875–1967). Babson made his fortune largely through the stock analysis and prediction company he founded in 1904, Babson's Statistical Organization (still extant today as Babson–United Inc.). His *Babson's Reports*, also founded in 1904, was the US's first investment advisory company, publishing one of the first investment newsletters. It was his theory that economies behave in accordance with Newton's third law of motion, that "to every action there is an equal and opposite reaction" – a belief which he would support by reminding people that, based on it, he'd predicted the Great Depression where others had not.

In 1940 he ran for the US Presidency on the Prohibition Party ticket, but succeeded in polling fewer votes than even the Socialist Party of America's Norman Thomas. He founded the Babson Institute (now Babson College) in 1919, Grace Webber College (with his first wife) in 1927, Utopia College (later called the Midwest Institute) in 1946, and the Open Church Foundation in 1947. But from our point of view his most notable achievement was the setting up in 1948 of the Gravity Research Foundation, a body specifically designed to discover means of reducing or entirely blocking the influence of gravity.

Roger Babson, the high priest of antigravity

His interest in – and detestation of – gravity can be traced to two family tragedies. The first occurred as early as 1893, when his 13-year-old sister Edith drowned at a neighbourhood swimming-hole. "Gravity . . . came up and seized her like a dragon and brought her to the bottom."* In a ghastly example of history repeating itself, almost on the anniversary of Edith's death Babson's grandson Michael, aged 17, died in a drowning accident in 1947. It was in the following year that Babson set up his Foundation to investigate gravity. The Foundation was based in New Boston, New Hampshire. In the final short chapter of the 1950 edition of his autobiography *Actions and Reactions* he outlines some of the Foundation's purposes. For example,

> Physicians and surgeons are becoming more and more interested in the relation between gravity and the physical conditions of individuals. Certain circulatory and other troubles are now recognized as directly due to gravity. . . . It is even being

* Roger W. Babson, "Gravity – Our Enemy Number One" (1948).

thought that there is a correlation between accidents and disease and the different phases of the moon, which, if so, means that our chances of getting hurt varies [*sic*] with changes in the gravity pull on our bodies. Yet no actual data exists [*sic*] as to this possible relationship and its allied complications. . . .

Therefore, one of the tasks of our Gravity Foundation is to collect from hospitals, insurance companies and physicians the day and, if possible, the hour of a fracture and learn how this time correlates with the phases of the moon. Not only does the pull of the moon and sun counteract (or relieve) at certain times the downward pull of the earth on an individual, but this same gravity may affect the temporary judgment or awareness of the individual. After ascertaining definite data on the above it must further be recognized that the variation of this gravity pull of the sun and the moon may affect the judgment of individuals differently according to their mental capacity and development. . . .

But Babson, for all his labelling of gravity as a deadly foe, was also keen to pursue possibilities of exploiting it, as he explained under the heading "The Harnessing of Gravity for Free Power":

Thousands of attempts have been made by earnest inventors to develop a machine to operate by gravity without the aid of any fuel or other power. These attempts have been unkindly called "perpetual motion" machines and have been a source of much unfair criticism. These machines, as a rule, work by levers and weights. None of these [has] been really satisfactory . . .

In order to keep in touch with these past and future gravity developments, I purchased control of Invention Incorporated of Washington D.C. This company has three investigators at the Patent Office at all times scanning through all of the patents issued from week to week. These investigators are constantly on the watch for any machine, alloy, chemical or formula which directly relates to the harnessing of gravity. The general impression is that gravity will not be harnessed until a partial insulator, reflector or absorber is discovered to develop a differential. It is further believed that this discovery will be accomplished through stumbling upon some alloy which will give the desired results. Hence, the Foundation is encouraging all engineers and chemists who work with alloys to be on the watch for such a discovery. It surely would be a great blessing to mankind.

And a little later he added something into which much might be read:

> The Foundation has purchased a small water power mill at New Boston, which is in perfect running condition although the horse power is small. This may be used for some secret experiments of an interesting nature.

In the 1960s Babson, seemingly in order to promote the Foundation's reputation, paid for stone monuments to be erected in the grounds of 13 educational establishments in various parts of the US; these colleges were also given endowments that were to be held in trust for a specific period before being released to finance antigravity research. Although the wording on the monuments was not identical, that on the rock at Tufts University in Massachusetts is typical:

THIS MONUMENT HAS BEEN

ERECTED BY THE

GRAVITY RESEARCH FOUNDATION
ROGER W. BABSON FOUNDER

IT IS TO REMIND STUDENTS OF
THE BLESSINGS FORTHCOMING
WHEN A SEMI-INSULATOR IS
DISCOVERED IN ORDER TO HARNESS
GRAVITY AS A FREE POWER
AND REDUCE AIRPLANE ACCIDENTS

1961

The Babson Foundation held seminars that attracted even some fairly eminent scientists – such as Igor Sikorsky (1889–1972), designer of the first successful helicopter (1939) – but of considerably more importance were the annual essay contests it sponsored, drawing papers from all over the world on gravity-related subjects; as the Foundation's focus slowly shifted from antigravity toward gravity in general, the essays began to assume considerable scientific merit. The contest continues today, with recent winners including Nobel laureates like Stephen Hawking and George F. Smoot.

After Babson's death in 1967 the Foundation slowly wound down, and it's now essentially defunct except for the contest, which George Rideout Jr, son of the original director, runs from the basement of his home.

Whether as a result of the Babson Foundation or not, the 1950s seems to have been the decade when optimism about the imminence of an antigravity breakthrough was running at its highest. In response to the *Zeitgeist*, the *New York Herald–Tribune* ran a series of three articles in late 1955 by the paper's Military and Aviation Editor, Ansel E. Talbert. Although Talbert finds room for a few more cautious voices, his assumption is that antigravity research is the wave of tomorrow – a wave that is surely, surely just about to break.

> The same type of scientific disagreement which occurred in connection with the first proposals to build the hydrogen bomb and an artificial earth satellite – now under construction – is in progress over anti-gravity research. Many scientists of repute are sure that gravity can be overcome in comparatively few years if sufficient resources are put behind the project. Others believe it may take a quarter of a century or more.

One of the enthusiasts Talbert interviewed was the distinguished avionicist Grover Loening (1888–1976), the first aeronautics graduate of any US university and the first engineer the Wright Brothers hired:

> I firmly believe that before long man will acquire the ability to

build an electro-magnetic contra-gravity mechanism that works. Much the same line of reasoning that enabled scientists to split up atomic structures also will enable them to learn the nature of gravitational attraction and ways to counter it.

Aircraft and associated companies actively pursuing the pot at the end of gravity's rainbow included, according to Talbert and other journalists of the time, Glenn L. Martin, Convair, Bell, Sikorsky, Lear, Clarke Electronics and Sperry–Rand. The first of these firms was in the process of building a new laboratory between Washington and Baltimore for the Martin-financed Research Institute for Advanced Study. Here a focus was the "implications for future gravity research in the 'United Field Theory' of the late Dr. Albert Einstein".

The most promising avenue to pursue, according to Talbert and most of his expert witnesses, was electro-gravitics, the notion that electromagnetism and gravitation were two aspects of a single force, and so one might be used to counteract the other. This is still a prime contention of serious antigravity researchers today. Talbert cites William P. Lear (1902–1978), founder of the aircraft company bearing his name, on the benefits of being able to create "electro-gravitational fields whose huge polarity can be controlled to cancel out gravity":

> All the . . . materials and human beings within these fields will be part of them. They will be adjustable so as to increase or decrease the weight of any object in its surroundings. They won't be affected by the earth's gravity or that of any celestial body. This means that if any person was in an anti-gravitational airplane or space ship that carried along its own gravitational field – no matter how fast you accelerated or changed course – your body wouldn't any more feel it than it now feels the speed of the earth.

Also widely reported in the mid-1950s was the existence of a Canadian (or joint Canadian–US) electrogravitics research venture called Project Magnet, supposedly founded in 1953, or perhaps 1950, under the leadership of distinguished scientist Wilbur B. Smith. In fact, this was set up in

1950* by the Canadian Department of Transport with the purpose of investigating UFO reports, and was headed by Wilbert B. Smith (1910–1962); he was a radio engineer and author of the unfinished, posthumously published *The New Science* (1964), which was apparently "Assembled . . . from data obtained from beings more advanced than we are" – so it must be right – and characterized by such revelations as

> Once Awareness has understood, through the application of the Quadrature Concept, the establishment of the first nine Parameters, the further application of this Concept yields three more Parameters which bring Matter and Energy as we know them into being. . . . The two higher Fabrics require much more sophisticated manipulations than our mathematics are capable of to show the relationships existing between the lower Parameters and Fabrics and the higher ones, and at our stage of evolvement we can reach them only through personal mental activity aided by hints and direction from others who have already passed this way.

One of those who reported on Smith's electrogravitics views was Michael Gladych, in "Spaceships that Conquer Gravity" for the July 1957 issue of *Mechanix Illustrated*. Gladych can hardly control his excitement when talking about the subject, and has to resort to using EVEN MORE CAPITALS than Smith would in *The New Science* – whole words and phrases rather than merely initial letters. Thus we discover that "AT LEAST 14 UNITED STATES UNIVERSITIES AND OTHER RESEARCH CENTERS ARE HARD AT WORK CRACKING THE GRAVITY BARRIER" and much more besides. In this context even quite sober ideas begin to seem deeply implausible: a bellow that "The new discovery means that before long we shall be able to switch gravity on and off as we do electricity and electromagnetism" serves to obscure the fact that the discovery in question is some perfectly respectable research done at the Princeton Institute for Advanced Study into subatomic particles.

The idea of Smith's that caught Gladych's attention was a logical continuation of the idea Lear had expressed to

---

* And shut down in 1953, a fact the post-1953 reports often omit.

Talbert, that "All the . . . materials . . . within these fields will be part of them". Assuming an antigravity ship (or "G-ship") could be constructed, there should be no problem with atmospheric friction despite the enormous velocities of which the G-ship would be purportedly capable. This would be because it wouldn't take much additional ingenuity to extend the antigravitational field to include a blanket of air around the vessel, which blanket would buffer it from the friction effects. In Gladych's terms this becomes

> To take the singe out of the friction heat, Canadian scientists headed by Wilbur [*sic*] B. Smith contemplate an ingenious ELECTRO-MAGNETIC device.

Another research avenue Gladych enthuses about is electromagnetic repulsion. He concedes that the "earth's magnetic field is TOO WEAK TO REPEL or PROPEL a G-ship made of ORDINARY METAL" but is not disheartened, because there's a simple solution: "By RE-ARRANGING THE ATOMIC STRUCTURE we could GREATLY INCREASE THE DIAMAGNETIC PROPERTIES of the G-ship . . ."

A rather more sober assessment than Gladych's of the situation had been prepared a few months earlier, in December 1956, at Gravity Rand Ltd of London in a document called "The Gravitics Situation":

> There are several promising approaches. One of them is the search for negative mass, a second is to find a relationship between gravity and heat, and a third is to find the link between gravitation and the coupled particles. Taking the first of these, negative mass, the initial task is to prove the existence of negative mass . . .

The Gravity Rand paper dwells on the work of Thomas Townsend Brown (1905–1985), on the Biefeld–Brown effect, and on the report Brown prepared for the US Government, Project Winterhaven. All three of these turn up repeatedly in the antigravity literature.

Thomas Townsend Brown was a US physicist whose career started conventionally enough. In 1921 he discovered what has become the most popular basis for electro-

gravitics research, the effect whereby passing a high voltage
through disparately shaped and sized electrodes can cause
a flow of charged particles (an ion wind) between them, the
impacts of those charged particles with air molecules gener-
ating a thrust on the apparatus as a whole. This has become
known as the Biefeld–Brown effect, because Brown was
working in the lab of Professor Paul Alfred Biefeld
(1867–1943) around this time; it is, however, unclear how
much if anything Biefeld contributed to the discovery.*
Brown was a firm believer in flying saucers, and became
convinced these must use the effect in their drive. Later
researchers have demonstrated that the effect doesn't work
in a vacuum, so it's not actually useful for a spaceship
drive; on the other hand, it's perfectly plausible our
ufonauts might use one drive in space and another in
atmospheres.†

Brown worked as a scientist for the US military and
then the US Government for a number of years before
becoming involved with a succession of commercial enterprises after the mid-1940s. He demonstrated electrogravitic
drives using disc-shaped models (the UFO influence!) bearing high electrostatic charges along their leading edge, and
in 1952 he presented to the US Government detailed plans
for a viable disc-shaped aircraft in a document called
Project Winterhaven. By the mid-1950s, as we've seen,

---

* Today the effect is perhaps more often called electrohydrodynamics.

† Some outlier results have suggested there might be some very small
remnant thrust even in the absence of atmosphere – that there's some
other process beyond ion flow at work in the Biefeld–Brown effect. In
space, where there's no such thing as air resistance and little gravity to
worry about, a small thrust can be useful: even the tiniest acceleration
can in time build up a respectable velocity. But a major problem for the
"ion wind drive" is the matter of power supply. In a desktop model, the
power supply can stay on the desk while the capacitor, connected by a
wire, eerily rises. Obviously that isn't feasible in space. Once lifted into
orbit by other means, you could perhaps add your power supply to your
interplanetary or interstellar spacecraft and take it with you, but, if you
could make your power source small enough and light enough for this
to be a sensible option, there are plenty of more effective ways to use it
for space propulsion.

several aircraft companies – as well as the US Government – were eagerly investigating the possibility of developing antigravity propulsion based on the Biefeld-Brown effect. By the later 1950s, however, this enthusiasm had ebbed sharply. According to antigravity enthusiasts this was because the US Government decided to classify the whole area of investigation once Brown's "gravitators" had achieved weight-reductions of some 30%; more likely the various organizations lost interest as it became clear the research wasn't going anywhere fast.

Matters weren't helped by the fact that Brown was intensely secretive about his work; good scientists aren't likely to believe reported results unless they're given at least a theoretical opportunity to replicate them. If the results aren't presented *at all*, just unsupported claims, obviously no replication is feasible. No wonder most scientists simply ignored Brown's work. However, in a November 1958 *FATE Magazine* article, "Townsend Brown and his Anti-Gravity Discs", one of Brown's defenders, Gaston Burridge, trumpeted this secrecy as a virtue:

> Since 1923 Brown and his family have spent nearly $250,000 of their own funds on experiments and research into . . . the "Biefeld–Brown Effect". Electrical literature contains few writings on this subject, mostly because Brown has maintained a tight grip on the information and has not seen fit to write on the matter scientifically or otherwise. No one else has seemed inclined to research the matter. What is more, American scientific journals are open to few idea that DO NOT ORIGINATE with men CONNECTED WITH LARGE UNIVERSITY or COMMERCIAL RESEARCH LABORATORIES!

The UK aeronautical engineer A.V. Cleaver, writing in the April/June 1957 issue of the *Journal of the British Interplanetary Society* ("Electro-Gravitics: What It Is – or Might Be"), echoed many when he insisted that what any proposed antigravity drive required wasn't just a new development from known principles but a brand-new, hitherto-unknown physical principle. A similar point had been made by Bryce DeWitt in an essay that won him the 1953 Babson Foundation award, "New Directions for Research in the Theory of Gravitation": rather than devote energies to investigating antigravity, it made more sense to put the effort into investigating the phenomenon of gravitation, about which far too little was known.

Far too little was known *by terrestrial science*, anyway! DeWitt must have blenched as ufologists and conspiracy theorists seized on the idea that antigravity devices were very much a matter of present reality, the US Government having back-engineered them from the crashed saucer retrieved in 1947 at Roswell, New Mexico.* Even if this weren't the case, it seemed obvious – from the speed of their manoeuvres – that flying saucers must rely on inertia-less drives, and what else could these be but antigravitic in nature? Rumours abounded and proliferated, were elaborated, proliferated again.

A principal figure was Major Donald E. Keyhoe (1897–1988). Keyhoe served as a US Navy aviator between the wars, spent a number of years as a freelance writer and as a writer, editor and PR man for the predecessor of the Federal Aviation Administration, was assigned as an aide to Charles Lindbergh after the latter's solo crossing of the Atlantic, wrote aviation adventure fiction and pop science articles for pulp magazines, was recalled as a training officer to the military during WWII with the rank of Major, then afterwards resumed his life as a freelance writer. In 1947 he was asked by *True Magazine* to investigate flying saucers, and became convinced this was no hoax or bout of mass hysteria. The result was the article "Flying Saucers are Real" (1950); the book *The Flying Saucers are Real* (1950) was a bestselling expansion of it. Further books followed, their recurring theme being that the USAF knew all about the saucers but was running a formidable coverup operation. In 1957 Keyhoe became Director of NICAP (National Investigations Committee on Aerial Phenomena), the saucer organization he'd cofounded with Thomas Townsend Brown the previous year.†

According to a 1967 article by Keyhoe ("Antigravity", in

---

* Somewhat later, such theories focused on the B-2 stealth bomber.

† Brown was NICAP's first director, but proved inept in financial matters and was ousted in Keyhoe's favour.

the magazine *Flying Saucer*) no fewer than 46 US Government projects were underway, involving

> experiments and research at two Air Force Laboratories (Flight Dynamics and General Physics Research), Radio Corporation of America, Massachusetts Institute of Technology and several technical engineering centers. In addition, official projects are being carried out at Barkley and Dexter Laboratories, Fitchburg, Massachusetts; Israel Institute of Technology; the Universities of California, Denver, Harvard, Indiana, Manchester (England), Maryland, Michigan, Minnesota, Ohio, Purdue, Stockholm (Sweden), Syracuse, Texas, and two New York schools – Queens College and Yeshiva Graduate School of Science.
>
> And, of course, some government agencies have projects so secret that they are not publicly registered and cannot be revealed without permission. . . .

Keyhoe and NICAP were at the forefront of the protests that the Air Force's Project Blue Book investigation of UFOs was just a coverup, and they made a number of representations on saucerish matters to Congress, but in all honesty their influence in the corridors of power was fairly negligible. Keyhoe's influence on the public, however, was for a while weightier, although it declined as the public either wised up or, possibly, succumbed to a certain amount of "saucer-flap exhaustion".

An inspiration for Keyhoe's views on antigravity, UFOs and much else were the views of the physicist and rocketry pioneer Hermann Oberth (1894–1989). Surprisingly, in view of the rest of his work, Oberth became a significant supporter of the idea that UFOs are alien spacecraft:

> I think that they possibly are manned by intelligent observers who are members of a race that may have been investigating our earth for centuries. I think that they have been sent out to conduct systematic, long-range investigations, first of men, animals and vegetation, and more recently of atomic centers, armaments and centers of armament production. They obviously have not come as invaders, but I believe their present mission may be one of scientific investigation.[*]

[*] Oberth, "Flying Saucers Come from a Distant World", *The American Weekly*, October 24 1954.

In his 1967 article on antigravity Keyhoe cited Oberth as saying to him, in 1961:

> With ordinary propulsion, such violent accelerations and maneuvers [as are reported of UFOs] would endanger the ship. Also, the force would crush any creatures aboard against the rear or sides of the machine. But with an artificial gravity field the force applies to the passengers and the spaceship. Even in swift changes of speed and direction, the ship is not strained and the passengers feel nothing.

Despite the stated opinions of such heavyweights as Oberth – and of the physicist Robert L. Forward (1932–2002), who theorized about a means of developing dipole gravitational fields that unfortunately relies on technology we do not yet possess – few scientists of Keyhoe's era showed much interest in developing an antigravity drive. Keyhoe believed many more scientists were interested in the subject than were willing to say so openly, either for fear of being laughed at, or, more sinisterly, because

> Air Force censors not only hide the facts but also belittle those who publicly report UFO sightings . . . But AF policy notwithstanding, the drive to get the secret of antigravity is well underway. It can't be stopped now. But it can be speeded up . . .

Well, the speeding up didn't happen, unless the conspiracy theorists are right and there's frenzied antigravitic research going on in Area 51 or thereabouts. There are, however, exponents of antigravitics still at work today. Probably the best known of recent researchers is Thomas E. Bearden (b1930), a retired US Army Lt-Col who is, according to his website, "Particularly known for his work establishing a theory of overunity electrical power systems,* scalar electromagnetic weapons, energetics weapons, and the use of time-as-energy in both power systems and the mind–body interaction."

Gosh. That's quite a lot to be particularly known for.

It seems Bearden first became interested in the idea of

---

* An overunity power system is one from which, in effect, you get more out than you put in. We have met many in this book.

overunity devices in the 1960s, when the story was going the rounds that the electrical engineers on the Minuteman missile, told to make the control circuitry as efficient as they could, inadvertently produced control circuitry that *was more than 100% efficient*: it actually generated an excess of electricity, which had to be got rid of before the final design could be used. This story has all the trappings of an urban myth; but clearly Bearden took it at face value and has never looked back.

His work most directly relevant to antigravity research was done in conjunction with Floyd "Sparky" Sweet and reported as "Utilizing Scalar Electromagnetics to Tap Vacuum Energy" (*Proceedings of the 26th Intersociety Energy Conversion Engineering Conference*, 1991). According to this paper, they managed, utilizing electromagnetic techniques, to reduce by 90% the weight of a 2.75kg object – an experiment that "supported a theory of gravitational force and antigravitational force". As far as I can understand, which is not far, the device is thought by Bearden to work because it creates "strong curves of local space-time that are local *strong negative gravity fields*".

This was far from Bearden's first foray into antigravitics, however. In 1981 he published a book called *Solutions to Tesla's Secrets and the Soviet Tesla Weapons*, in which he claimed to have discovered the nature of the electromagnetic-style waves that, in his turn, Nikola Tesla had claimed to have discovered decades earlier which travelled at many times the speed of light and whose signal strength did not fall off with distance. The amazing "Tesla waves" can, according to Bearden's book,

❖ ESTABLISH STANDING WAVES
  – in the earth
  – in the ionosphere
❖ TAP ENERGY FROM THE EARTH'S CORE
❖ TRAVEL FASTER OR SLOWER THAN LIGHT
❖ CHANGE RATE OF TIME FLOW
❖ AFFECT ALL FIELDS, INCLUDING GRAVITY
❖ COMMUTE BETWEEN VIRTUAL AND
  OBSERVABLE

It's clear this wasn't just one of those run-of-the-mill scientific breakthroughs that improve car mileage or increase computing speeds – although Tesla waves may well do all of these things and more. The USSR was, as the book's title implies, developing Tesla-wave technology for purposes of weaponry,* but

> The potential peaceful implications of Tesla waves are also enormous. By utilizing the "time squeeze" effect, one can get antigravity, materialization and dematerialization, transmutation, and mindboggling medical benefits. One can also get subluminal and superluminal communication, see through the earth and through the ocean, etc. The new view of phi-field also provides a unified field theory, higher orders of reality, and a new super-relativity, but detailing these possibilities must wait for another book.
>
> With two cerebral brain halves, the human being also has a Tesla scalar interferometer between his ears.

I have to confess that it was at this point in reading *Solutions to Tesla's Secrets* that my Tesla scalar interferometer refused to cooperate any longer.

Bearden's *magnum opus* is *Energy from the Vacuum* (2002), a book almost a thousand pages thick that Martin Gardner, reviewing it in *Skeptical Inquirer* in early 2007, described as "much funnier, for instance, than Frank Tipler's best-seller . . . *The Physics of Immortality*". Gardner expressed surprise about much within the book, including Bearden's support for the Dean Drive (see page 218) and his explanation of Wilhelm Reich's spurious orgone energy as "really the transduction of the time-polarized photon energy into normal photon energy" – whatever that might mean. As the book's title might suggest, Bearden is also a great enthusiast of ZPE, or zero-point energy, another name for the vacuum farming espoused by Harold Puthoff and others (see page 206).

Bearden's biggest claim to fame is the invention (or co-invention) of a perpetual motion machine, the Motionless

* And those pesky Russkies are, according to Bearden, so venal that they just won't quit: they used their Tesla-wave energy beams to shoot down the Space Shuttles in the disasters of 1986 and 2003.

Electromagnetic Generator, or MEG, a sort of jazzed-up transformer with a permanent magnet at its core. Initially external power must be supplied from a battery or the like, but once the device gets going it generates sufficient voltage to power itself, which it does thereafter. Bearden – with Stephen L. Patrick, James C. Hayes, James L. Kenny and Kenneth D. Moore – obtained a US patent for the MEG in 2002. Quite how they managed to do so is a mystery. As noted on page 205, the US Patent Office won't grant a patent to anything that seems to be a perpetual motion machine unless the applicants supply a working prototype; to date Bearden and his colleagues have failed to produce this.* The explanation sometimes offered for the granting of the patent is that it might not have been evident to the investigators just quite what the MEG was claimed to be, yet the opening line of the "Summary of the Invention" section in the patent application reads:

> It is a first objective of the present invention to provide a magnetic generator [in] which a need for an external power source during operation of the generator is eliminated.

This seems pretty clear.

Bearden had high hopes for the MEG, claiming in 2001 that within a year or so the first MEG-based products would be on the market. Others who have tested models based on the design given in the patent application have failed to achieve the desired results. Nonetheless, it seems investors have spent at least millions towards the further development of the device.

To criticisms that overunity devices are in violation of the known laws of physics, Bearden has the response that the MEG is far from the first overunity device to have been invented. Rather, governments and energy corporations have for over a century conspired to suppress all such developments – an example being cold fusion. This was claimed in 1989 by Stanley Pons and Martin Fleischmann of the University of Utah, but relatively soon discredited.

---

* In 2005 Bearden claimed they had produced a working prototype, but that it had been destroyed.

According to Bearden, what really happened is that it was suppressed and its discoverers smeared by the nuclear power-plant industry. When one of cold fusion's stalwarts, Eugene F. Mallove (1947–2004), editor of the magazine *Infinite Energy*, was beaten to death, ostensibly by a pair of crack-heads intent on burglary, it was obvious to Bearden the nuclear industry was behind the hit. Researchers into overunity devices are particular targets for covert assassination; Bearden devotes nearly fifty pages of *Energy from the Vacuum* to the persecution of fringe scientists by the agents of secret forces, and estimates that as many as fifty of those murdered have been overunity researchers. He himself has several times been the victim, but luckily never fatally so, of "shooters", devices that fire projectiles that simulate heart attacks in the victims. Another technique favoured by the Forces of Darkness is to ram a researcher's car from behind, then, in the ambulance that just happened to be passing, turn him into a human vegetable by injecting air into his bloodstream.

Overunity, Tesla rays and cold fusion are not the only energy technologies to have been suppressed or kept secret. The most alarming owners of such supertechnologies must be the Yakuza. They have been part of conspiracies to use supertechnology to create, among much else, the 2004 tsunami and the following year's Hurricane Katrina. Included in the Yakuza's future plans is the detonation of the supervolcano under Yellowstone Park, an eruption that would inevitably – and here we intersect briefly with orthodox science – lead to the annihilation of much of the life (humans included) on the North American continent.

Bearden does not restrict himself entirely to physics. He has also developed a cure for cancer: as far as I can work it out, all you have to do is send the affected cells back in time far enough that they're healthy again. This seems to be a part of his overall interest in "Time Engineering". The beginning of his description on his website of his achievements in this field reads:

> Proposed a mechanism that generates the flow of time itself, so that the overall time flow is thus engineerable and also

possesses a rich internal structure and dynamics of other internal streams of the flow of time.

A thinker of similar polymathic spread is Noel Huntley PhD.* In his paper "Antigravity and the Ultimate Spacecraft Propulsion System" Huntley offers not one but three different approaches to the antigravity conundrum.

The first of these depends on the fact that the "positively charged nucleus at the centre of the atom is a white hole". As such, the nucleus is a portal whereby energy is streaming from elsewhere into our universe (or our "3rd dimension", as Huntley calls it). All is normally in equilibrium, because electrons act much like teeny black holes: energy vanishes *via* them from our universe back into the elsewhere. The trick to achieve antigravity is to ionize (strip away the electrons of) the atoms of one of the inert gases to leave only those white-hole nuclei.† Place the whole body of ionized gas under the influence of a powerful magnetic field and it's obvious the enormous inflow of arcane energy will produce a force that'll repel any matter it comes up against. For your antigravity drive, then, all you do is harness this force appropriately – a trick Huntley concedes he has not himself yet done, although "This system has apparently been achieved secretly, and since we haven't heard about it publicly we can only assume it is the work of the secret government. But they have more advanced systems than this now."

The second technique Huntley describes relies upon in effect reversing the direction of the local gravitational field. Here Huntley points us towards the work of two pioneers, John Searle in the UK and Otis Carr in the US. It seems Searle was able to give his home a free electrical supply

* It seems you can diagnose the unorthodoxy levels of people's physics by the aggressive appending of "PhD" to their names. Bearden used to do this too until it was revealed his PhD came from a degree mill called Trinity and University College that operates out of PO boxes in various countries. This institution granted him his PhD for "Life Experience".

† The choice of an inert gas is so things won't get complicated by chemical reactions.

> Several years ago, Glenn Martin's vice-president for
> advanced design, G.S. Trimble, predicted that by 1985
> practically all airliners would be using artificial gravity,
> flying at almost unbelievable speeds.
> – Donald E. Keyhoe, "Antigravity",
> *Flying Saucer Magazine*, 1967

using a generator he'd built based on the principle (but
then the authorities stepped in and confiscated the device)
while Carr built a full-scale working spacecraft (but then the
authorities – stop me if you've heard this one – stepped in
and confiscated the device). According to Huntley:

> If one envisages a doughnut-shaped energy configuration
> encompassing the disc, which is a scalar field, for a particular
> rotation, gravitational nodes descending vertically under the
> force of gravity would be swept around the disc and then
> underneath and upwards and round again, delineating and
> penetrating the vortex region or toroid shape. With the same
> principle as the twister and tornado, which can lift a vehicle,
> the gravitational nodes coming up under the disc, and
> resonating with the nodes of atomic structure of the disc,
> would drag the levity disc upwards. Thus in this example grav-
> ity is reversed. The spacecraft "falls" up.

The third of Huntley's means of developing an antigravity
device is somewhat more hypothetical, since it depends
upon your being able to get hold of some of the material
that extraterrestrials use to build their flying saucers – some
"of [whose] advanced designs are apparently created in
space using the psychokinetic powers of several minds".

With all of the bile directed towards stuffy old Einstein
by the antigravity researchers and the proponents of
perpetual motion and free-energy devices, it's almost
refreshing to find that William Lyne, in his pleasingly anar-
chic *Occult Science Dictatorship* (2002), sets a pox on *both* their
houses. Indeed, according to Lyne they're really one and
the same:

> Today, suddenly, masquerading as "free-energy researchers",
> there appears a plethora of Relativists. Relativists are like

broken records, and cannot get out of the same old Relativist "groove". If you look at their websites, and read their material, all you see is the "same old same old" contorted Relativist "theories". They theorize, and theorize, and theorize, but what do they really do? They theorize . . . curved space . . . dilated time . . . and ruminate through all the other aspects of the shaky "house of cards" built by St. Albert. And out of these insane ruminations, they attempt to solve the riddle of "gravity", "anti-gravity" and "fluctuations in the quantum vacuum field", but no solutions are in sight now, just as none were in sight in 1905.

Lyne himself is a staunch supporter of the luminiferous aether, the universal medium which it was once supposed electromagnetic radiation required in order to travel from one place to another. In Lyne's view, the aether is not the same in all parts of the cosmos, but tends to be "rigidified" in the proximity of massive celestial bodies. The red shift observed in the spectra of distant galaxies, supposed by most to be a sign that the universe is expanding, is really a refraction effect produced by rigidified aether. There was no such event as the Big Bang, and the universe is eternal – although this situation is *not* that predicted by the steady state theory of Gold, Bondi and Hoyle, which for some reason is every bit as vile as Big Bang/expanding universe cosmologies.

In 1996 the teenaged UK hacker and UFO-enthusiast Mathew Bevan was arrested in connection with his successful penetration of US military computers; one Pentagon spokesman described him as "the biggest threat to world peace since Adolf Hitler".* In a 1998 interview with Matthew Williams, Bevan claimed he'd discovered details of an antigravity propulsion unit in the computers of the Wright–Patterson Air Force Base, Ohio:

> . . . there was one machine on the network where I read current files and future project proposals. I read documents which gave the impression that they had an anti-gravity engine

---

* So much for hyperbole. In late 1997 the UK's Crown Prosecution Service decided it wasn't in the public interest to pursue the case against Bevan, and dropped it. Adolf Hitler wasn't so lucky.

which was capable of at least Mach 12 to Mach 15. . . . Supposedly the aircraft which employs this engine uses a reactor which there were a lot of detailed numbers and figures for, but I have no idea what all this meant. I can remember that the documents referred to a super heavy element, whatever that means. The element is the main fuel for the reactor. The engine worked by making a disturbance of molecules at the front of the craft so that it was able to stop the inertia or G-force inside the craft. I got the impression that this information was the type of material I was looking for because it was far in advance of our current technology and could be something to do with the Roswell UFO. . . .

. . . [T]he interviewing officer asked me if I knew what Hangar 18 meant. I said, "Well, if you are thinking of a building where they store extraterrestrial aircraft . . . is this what you mean?" He replied that this could be the place that he was thinking of. This was the only time that Hangar 18 was mentioned in the interview.

No discussion of antigravity would be complete without a brief look at the strange case of Edward Leedskalnin (1887–1951). Born in Latvia, Leedskalnin came to North America *c*1915, eventually, after a bout with tuberculosis, settling in Florida City, Florida. There he began building a monument called Rock Gate, which he dedicated to a girl who had, back in Latvia, jilted him. What was exceptional about Rock Gate Park, which he lived in while constructing it, was that it was constructed out of massive blocks of solid coral – some 1100 tonnes of it, all told.

No one knows for sure how this slight man could, on his own and without evidence of heavy machinery, have hewn and manoeuvred the weighty slabs; and even less can people understand how he managed, in the mid-1930s, to uproot the entire monument and transport it to a new home near Homestead, Florida (where it is now known as the Coral Castle). Inevitably, many people have found the temptation irresistible to claim Leedskalnin must have invented some kind of supertechnology: specifically, an antigravity device.

Such a hypothesis is barely quenched by the "perpetual motion holder" Leedskalnin described in three (of four) esoteric pamphlets he published: *Mineral, Vegetable and Animal Life* (1945), *Magnetic Current* (1945) and *Magnetic Base* (n.d., but likely 1945). The device was a horseshoe magnet whose ends were linked by a piece of iron wrapped in a wire coil; if a jolt of electricity were applied to the coil, electricity would continue to circulate around the closed loop of magnet and iron bar until the two were separated, at which point, even if this were months later, the electricity could be reclaimed from the coil. Leedskalnin made astonishing claims for this gadget in the pamphlets – and for magnets and magnetism:

> Magnets in general are indestructible. For instance you can burn wood or flesh. You can destroy the body, but you cannot destroy the magnets that held together the body. They go somewhere else. [*Mineral, Vegetable and Animal Life*]

> As I said in the beginning, the North and South Pole magnets they are the cosmic force. They hold together this earth and everything on it, and they hold together the moon, too. . . . Here is a good tip to the rocket people. Make the rocket's head strong North Pole magnet, and the tail end strong South Pole magnet, and then shut to on the moon's North end, then you will have better success. [*Magnetic Current*]

As will be evident, it's not always possible to be confident of what Leedskalnin's talking about, and this leaves the door wide open for unorthodox physicists and conspiracy theorists. One possibility that has been touted is that the texts of the pamphlets, with their strangely stilted prose style (which others might attribute to Leedskalnin's first language being Latvian), contain, in encoded form, the *real* information about Leedskalnin's electromagnetic researches. The moment seems ripe for a Dan Brown novel on the subject.

The Orffyrean Wheel, the Kenyon Alternator, the Garabed, the Keely motor, Podkletnov's antigravity shield, even Congreve's sponges . . . start googling for them and you'll soon find they all still have their advocates. It's only, you'll read, because of political conspiracies and/or the corruption of the self-serving scientific Establishment that these devices haven't been permitted to transform the course of human civilization and bring peace and plenty to the world.

There is also, of course, the problem that they don't work.

What sort of a mindset imagines scientists would conspire to suppress devices that could produce such over-whelming good? Established industrial corporations, yes – so bad is their track record it's easy to believe they might stomp any device that threatened their current business model. But the world of physics as a whole? In his 1998 article "Breaking the Law of Gravity" (*Wired* #6.03) Charles Platt cited Ron Koczor of NASA on the potential wonders opened up by the Podkletnov research:

> But if this is real, it's going to change civilization. The payoff boggles the mind. Theories about gravitational force today are probably comparable to knowledge of electromagnetism a century ago. If you think what electricity has done for us since then, you see what controlling gravity might do for us in the future.

That doesn't sound much like an Establishment scientist hushing up an inconvenient technological development.

Some people really believe there are undiscovered monsters at loose in the world. Take, for example, werewolves.

In his *The Book of Were-wolves* (1865), Sabine Baring-Gould (1834–1924) quotes *in extenso* a 1508 sermon by the Strasbourg preacher Johann Geiler von Keysersperg (1445–1510) on the subject. Interestingly, for the most part von Keysersperg seems to regard werewolves as normal wolves that have become imbued with a lust for human flesh rather than as men who transform into wolfish form in order to perpetrate acts of savagery. He lists seven reasons

why wolves may become maneaters, of which the first five
– hunger, savagery, old age, experience and madness –
perhaps surprisingly for the period invoke no supernatural
factor. The sixth and seventh, however, are of a different
order:

> Under the sixth head, the injury comes of the Devil, who
> transforms himself, and takes on him the form of a wolf. So
> writes Vincentius in his *Speculum Historiale*. And he has taken it
> from Valerius Maximus. In the Punic War, when the Romans
> fought against the men of Africa, when the captain lay asleep,
> there came a wolf and drew his sword, and carried it off. That
> was the Devil in a wolf's form. The like writes William of
> Paris,—that a wolf will kill and devour children, and do the
> greatest mischief. There was a man who had the phantasy that
> he himself was a wolf. And afterwards he was found lying in the
> wood, and he was dead out of sheer hunger.
>
> Under the seventh head, the injury comes of God's
> ordinance. For God will sometimes punish certain lands and
> villages with wolves. So we read of Elisha,—that when Elisha
> wanted to go up a mountain out of Jericho, some naughty boys
> made a mock of him and said, "O bald head, step up! O glossy
> pate, step up!" What happened? He cursed them. Then came
> two bears out of the desert and tore about forty-two of the
> children. That was God's ordinance. The like we read of a
> prophet who would set at naught the commands he had
> received of God, for he was persuaded to eat bread at the
> house of another. As he went home he rode upon his ass. Then
> came a lion which slew him and left the ass alone. That was
> God's ordinance. Therefore must man turn to God when He
> brings wild beasts to do him a mischief: which same brutes
> may He not bring now or evermore. Amen.

Myth? Obviously. Yet there's an infuriatingly fuzzy border
when considering cryptozoology – the science or pseudo-
science, take your pick, of unknown animals – between myth
and sober rationality.

It cannot reasonably be argued that the existence of
as-yet-undiscovered large terrestrial animals is an impossibility, or even a particularly strong implausibility; that there
are large *marine* animals we don't know about seems a fairly
safe bet. A good example of how we know far less than we
sometimes think we do about these matters concerns the

giant sloths – species of *Megatherium* and *Mylodon*. When
reports reached Europe in the early 18th century of bones
the size of elephant bones being unearthed in South
America, no one believed the stories: everybody knew there
weren't elephants in South America. In 1789, however, a
complete skeleton was found; this, the first fossil skeleton to
be identified as mammalian, was sent to Madrid and
mounted by the dissector Juan Bautista Bru (1740–1799).
In 1796 the anatomist Jose Garriga published a description
of the creature, which was given the name *Megatherium
americanum*. Over succeeding decades it became clear there
had been several species of these giant sloths; it was,
however, assumed they must have been long extinct before
human immigrants from Asia came on the scene. Then a
Megatherium skeleton was discovered where it was evident
the creature had been trapped in a pit and roasted alive
there – a clear sign of human activity. The clincher came
some while later with the discovery in a Patagonian cave of
a human skeleton and the remains of two *Mylodon* its owner
and his friends had clearly been eating.

Before this discovery there was a reputable sighting in
Patagonia of a creature that seemed remarkably like one of
the smaller *Mylodon* species. Certainly the Argentinian naturalist Florentino Ameghino (1854–1911) came to think so.
Whether he was correct is still a matter of debate – especially since a passage in the writings of the French priest and
explorer André de Thévet (1502–1590) that describes a
beast in Patagonia called the *su*, and the way in which the
Indians hunted it, seems yet again to indicate a variety of
Mylodon. If the giant sloth was being hunted in the late
Middle Ages, how can we be sure there aren't still survivors
in a part of the world where there remain plenty of untamed
areas? Sir Ray Lankester (1847–1929), Director of London's
Natural History Museum, thought there was every
possibility this could be so, and other zoologists have
echoed his view.

Zoologists and palaeontologists can be deluded by
fakes; but conversely they can erroneously deny outright
(rather than merely doubt) the existence of a creature
simply because unwilling to accept the evidence of witnesses

from outside their own field. A prime example of the latter phenomenon is the case of the coelacanth. In 1938 some fishermen off the coast of East Africa hauled one of these fish out of the sea, still thriving 70 million years after its supposed extinction. Further questioning of the region's fishermen elicited the information that they were perfectly familiar with the coelacanth, *Latimeria chalumnae*, and had never realized there was supposed to be anything interesting about it: no zoologist had thought to ask them.

And hitherto unknown species still turn up all the time. For example, in 2005 an entirely new type of creature was discovered, *Kiwa hirsuta*, which lives around some of the Pacific deep-sea hydrothermal vents in an environment that would be toxic to most other animals. *Kiwa hirsuta* is lobster-like, about 15cm long, and "furry" – it has tufts of quite long hair-like strands on its main pincers and here and there on the body. The creature is so distinct that the taxonomists created a whole new family for it, Kiwaida: there may be rafts of other Kiwaida genera waiting to be discovered. Zoologists are far less sanguine than they used to be, even a couple of decades ago, that the major areas of their map of life on earth have all been shaded in. Even new species of large mammals have been discovered in the recent past.

To be fair to zoologists and paleontologists, we should add that they commit far fewer howlers than do the hordes of amateur enthusiasts whose help is far too generously offered. Consider the many amateur sightings of bigfoot, or the Loch Ness Monster, and the photographs of both that have been revealed eventually as fakes. Are the quasi-scientific explanations of the Loch Ness Monster – that it's a family of plesiosaurs, or some similar dinosaur – any more plausible than the suggestion in the Open University Geological Society's *Journal* in 2006 by Neil Clark, Curator of Palaeontology at Glasgow's Hunterian Museum, that what people have seen when they think they've seen Nessie are elephants, allowed out for a swim by travelling circuses on their way to and from Inverness? "When their elephants were allowed to swim in the loch, only the trunk and two humps could be seen – the first hump being the top of the head and the second being the back of the animal." Both are hypotheses. Both draw upon the implausible – in one instance that a dinosaur could survive a few tens of millions of years after the extinction of the rest without its presence being considerably more noticeable than it is, in the other that observers wouldn't notice crowds of circus personnel on the shoreline yelling at Nellie to come back right this minute. Or perhaps Clark was joking.

The zoologist Karl P.N. Shuker (b1959), in his book *Mystery Cats of the World* (1989), called upon the shade of Sherlock Holmes in declaring:

> After having eliminated the impossible relative to mystery cats – viz. that all reports of all such creatures result from dimly viewed dogs, manic mendacity, drunken delusion or mass hallucination – we are left with an initially improbable but ultimately inevitable conclusion: namely, that mystery cats of wide diversity and worldwide distribution do exist.

The zoologist John Napier (1917–1987) made much the same observation of the North American bigfoot and its cousins around the world. The zoologist Bernard Heuvelmans (1916–2001) made it of not just the bigfoot but of sea serpents. Today the anthropologist Jeff Meldrum (b1958) makes it very forcefully about bigfoot. Even if the search may ultimately prove fruitless, it's worthwhile. While some of the monsters sought by amateur cryptozoologists really are fanciful,* and some of the claims even more so, there is no *theoretical* reason for dismissing notions of the existence of creatures like bigfoot and the sea serpents. The fact that cryptozoology has attracted more than its fair share of fakers, hoaxers and bogus scientists should not obscure this point.

Unidentified large wild cats are popular cryptozoological sightings. In the UK the Exmoor Beast and the Surrey Puma frequently make the headlines; it seems overwhelmingly probable these cases represent misperceptions or exotic escapees. In North America there's the black panther,

---

* For example, according to our old friend Sylvia Browne in her *Secrets & Mysteries of the World*, sasquatch/bigfoot, the Loch Ness Monster and even the mythological Leviathan are all tulpas!

What monsters might the oceans hold?

different sightings of which majestic creature may more prosaically be of the Central American jaguarundi wandering well outside its accepted range; or of survivals of the black jaguar, known to have been native, albeit rare, to parts of the US as recently as the early 20th century; and the fisher, an animal of the marten family that while not in fact feline can seem so and is native through the northeastern US and much of Canada. Other reputed North American mystery cats include the amphibious wampus of Missouri and the whistling wampus of Arkansas, the sap-drinking cactus cats of Arizona, the vast wowzer (like a puma but very much bigger) of Oklahoma and the santer of North Carolina (although some accounts say this is more doglike than catlike). There are also occasional reports of lions being seen – African-style lions with manes, not mountain lions. These are likely escapees from zoos and circuses or, revoltingly, from illegal big-game shoots using imported animals.

In Australia the marsupial equivalent of the big cat was the Tasmanian tiger or thylacine, *Thylacinus cynocephalus*, which looked rather like a wolf (hence its alternative name of Tasmanian wolf) but had tigerlike stripes on its lower body. This creature was once widespread throughout Australia but lost out in the ecological competition to the dingo except in Tasmania, where it was eventually hunted to extinction by the white man. Occasional reports of sightings of animals that fit this description suggest a small population may still survive unknown in one of the island's remoter areas. It's tempting to think the Queensland tiger, whose zoological status is considerably less sound, might be of the same stock; but reports say this has a typically feline face whereas the thylacine had/has a lupine-looking face. During the 19th and early 20th centuries these reports were frequent enough that it seems not to have occurred to anyone that people were observing anything other than an element of the local wildlife, which of course is what the Queensland tiger may well have been. One possibility is that these creatures may have been survivals of the marsupial lion, *Thylacoleo carnifex*, supposedly extinct 10,000 years ago.

In Argentina the iemisch, or hymché, bears many resemblances to the tiger but has webbed feet, its lifestyle being largely aquatic. In 1897 the noted Argentine palaeontologist Florentino Ameghino received an account and a few small supposedly iemisch bones from his brother Carlos. The latter wrote:

> This animal is of nocturnal habits, and is said to be so strong it can seize horses with its claws and drag them to the bottom of the water. According to the description I have been given, it has a short head, big canine teeth, and no external ears; its feet are short and plantigrade, with three toes on the forefeet and four on the hind; these toes are joined by a membrane for swimming, and are also armed with formidable claws. Its tail is long, flat and prehensile. Its body is covered with short hair, coarse and stiff, of a uniform bay colour. In size it is said to be larger than a puma, but its paws are shorter and its body thicker.

Florentino was so convinced this must be an example of the giant sloth *Megatherium* (see page 246) that, when he wrote of the iemisch himself, he altered the number of toes the beast was supposed to have. Most cryptozoologists think either the iemisch is/was some unknown species of giant otter or that the accounts were hugely inflated tales conflating two creatures known to be in the region: the otter, which matches the aquatic criterion, and the jaguar, a suitably fierce large quadruped.

Whatever the status of the kraken, real or legendary, sober accounts of these monsters exist. In 1680, for example, a supposed kraken managed to strand itself on the Norwegian coast; for months, as the huge carcase decayed, the stench was so strong no one could go within miles of it. The best known kraken account is probably that of the Danish missionary Hans Egede (1686–1758), Bishop of Iceland, in his book *Det Gamle Grønlands nye Perlustration* (1741; *A Description of Greenland*):

> As for other Sea Monsters . . . none of them have been seen by us . . . save that most dreadful Monster, that showed itself upon the Surface of the Water in the year 1734, off our colony in 64 degrees. The Monster was of so huge a Size, that coming out of the Water its Head reached as high as the Mast-Head; its body was as bulky as the Ship, and three or four times as

Bing's drawing of the kraken seen by Egede

long. It had a long pointed Snout, and spouted like a Whale-Fish; great broad Paws, and the Body seemed covered with shell-work, its skin very rugged and uneven. The under Part of its Body was shaped like an enormous huge Serpent, and when it dived again under Water, it plunged backwards into the Sea and so raised its Tail aloft, which seemed a whole Ship's Length distant from the bulkiest part of its Body.

The creature wasn't a whale – there was plenty of whaling going on at the time in these waters, so even if Egede didn't recognize a whale the crew certainly would. Modern writers tend to believe it must have been a giant squid. Whatever, Egede asked a preacher called Bing to draw a picture of the creature he'd seen, and this picture, reproduced in Egede's book, is one of the earliest known of a sea monster.

Another Scandinavian bishop, Erik Pontoppidan (1698–1764), Bishop of Bergen, recorded in *Versuch einer Natürlichen Geschichte Norwegens* (1752–3; *The Natural History of Norway*) numerous sightings of the kraken reported to him by Norwegian fishermen. It's difficult to know how seriously to take these, for in his other books Pontoppidan revealed a broad streak of credulity. However, for what it's worth, he described the kraken thus:

> Its back or upper part, which seems to be about an English mile and a half in circumference, looks at first like a number of small islands surrounded with something that floats . . . like seaweed. . . . At last several bright points or horns appear, which grow thicker and thicker the higher they rise above the surface of the water, and sometimes they stand as high and large as the masts of middle-sized vessels. It seems these are the creature's arms and, it is said, if they were to lay

A sea serpent described by Pontoppidan

hold of the largest man-o'-war, they would pull it down to the bottom. After this monster has been on the surface of the water a short time, it begins slowly to sink again, and then the danger is as great as before, because the motion of this sinking causes such a swell in the sea, and such an eddy or whirlpool, that it draws down everything with it.

The fishermen who reported these sightings to him were, far from living in dread of the monster, grateful to it. The presence of the kraken below the surface was likely to set the fish jumping, so the men could soon fill their nets. It was only if their soundings showed the kraken was preparing to surface that they knew to get out of the way as swiftly as their oars could take them.

Pontoppidan's contemporaries laughed his account out of court as just another example of a well meaning scholar being fooled by tall tales, but modern naturalists are a little more charitable, suggesting that what the fishermen might have encountered (and only moderately exaggerated) could have been a large group of feeding squid; on surfacing together they could give the appearance of "a number of small islands". To an extent this hypothesis is bolstered by the fact that the creatures the fishermen (and Pontoppidan) described as "baby kraken" have been fairly confidently identified as squid.

Some sightings of sea serpents may in fact be of snakes. The African python, which grows up to well over 6m long, reportedly up to 9.72m, and can swallow a goat, has on occasion been sighted swimming in the Indian Ocean, seemingly travelling from one island to the next in search of food.

The astonishing persistence of the zoological hoax can be exemplified by the supposed natural history of the gold-digging ants of India, a tall tale that was embellished over centuries as more "facts" became available. Strabo (c63BC–cAD21), in his *Geographia*, was possibly the first to put the matter down in writing, although he refers to earlier authorities. The information was picked up by Pliny the

Elder (23–79), who soberly recorded an account of the lifestyle of these creatures in his *Historia Naturalis*:

> In the northern part of India dwell ants which have the colour of cats; they are of the size of the Egyptian wolf. They grub gold from the ground. This they amass in the winter; in the summer they hide under the ground to escape the heat, and that is when the Indians steal the gold. But the Indians must be quick about it, for on smelling humans the ants come out of their holes, pursue the thieves and, if the men's camels are not swift enough, tear the men to pieces. Such is the speed and savagery the love of gold awakens in the ants.

Pliny's account seems to have been part of a broader effort – which likewise persisted for centuries – to claim evidence for the notion that animals, not just humans, had an insatiable craving for gold. This effort may have been a search for an excuse for humanity's own greed for the metal, or an attempt to demonstrate there was something special about gold that made it irresistibly desirable. Whatever, the gold-digging ants proved a popular addition to the bestiary.

In *Li Livres dou Trésor* (*c*1266) Brunetto Latini (*c*1210–*c*1295) enormously amplified the description of the creatures, while simultaneously shifting their homeland from India to an Ethiopian island and imbuing them with exceptional intelligence; their savagery remained undiminished. He gave considerable detail about how the Ethiopians managed to get the gold from the ants without being killed in the process. Presumably basing his account on Latini's, the German theologian and cosmographer Sebastian Münster (1489–1522) included an illustrated description of the gold-digging ant in his *Cosmographia Universa* (1544). The French historian Jacques Auguste de Thou (1553–1617) told how in 1559 the Shah of Persia included a gold-digging ant among gifts sent to the Sultan Suleiman. It was a savage creature about the size of a large dog.

It took well over 1500 years for natural historians to perceive that there must be something . . . well, *wrong* about these accounts. However, it went against their instincts simply to laugh the whole business off as a travellers' tale. Instead various hypotheses were advanced in an effort to demonstrate an underlying kernel of truth. Some pointed toward the burrowing engaged in by the Siberian fox: surely such an intelligent creature must have some purpose for all its digging, and what better purpose than to search for gold?

In this way the reasoning came neatly round in a circle. The legend was born out of the desire to show that animals coveted gold and would dig for it; now an animal's entirely hypothetical yen for gold was being taken as the justification for its digging habits. The phenomenon responsible for the fable's survival was what we can call belief perseverance: the irrational retention of a belief long after all the evidence shows it to be untenable. It's a phenomenon widely observed within the pseudosciences: where the facts contradict the belief, the facts are dismissed rather than the belief modified or discarded. There are countless other examples of belief perseverance in this book.*

While the idealized mermaid has been for the past 400 years or so a beautiful, nubile (albeit fish-tailed) maiden whose main occupation seems to be sitting on rocks and combing her locks, even the most credulous have always recognized that real mermaids can't be *quite* like that. Thus all any money-hungry hoaxer has had to do has been to produce a figure that is vaguely human at one end and (the easy bit) fishlike at the other. As with most fakes, dead "mermaids" were concocted from bits of other animals – or worse: in the 18th century a Dr John Parsons (1705–1770) discovered that a small "mermaid" on exhibit in London was in part composed of a human foetus.

* I was widely rebuked by right-wingers for using as an example of belief perseverance in my book *Discarded Science* (2006) the astonishing retention of belief by sections of US society in Saddam Hussein's WMDs long after their existence had been comprehensively disproved: this was, apparently, an attack on conservative views. Huh? In fact, of course, the rebukes merely served as further demonstrations of the power of belief perseverance: it is obviously neither conservative nor liberal to claim as fact something that has been shown false.

A late-19th-century mermaid fake from Japan

The most famous of all these monstrosities was brought to London in 1822 by a Bostonian called Captain Eades; it had been shown around the world for a few years before that, and reports from abroad ensured its arrival would be a *cause célèbre*. Eades himself seems to have been convinced of the authenticity of his grisly possession; certainly he showed no inhibitions in requesting Sir Everard Home (1756–1832), President of the Royal College of Surgeons, to examine the object. Home sent William Clift (1775–1849), Curator of the Hunterian Museum, in his stead. Clift reported:

> The cranium appears evidently to belong to an orang-outang of full growth, the teeth, and probably the jaws, do not belong to the cranium, but from the size and length of the canine teeth, they appear to be those of a large baboon. The scalp is thinly and partially covered with dark-coloured hair, which is glossy like that of an orang-outang. The skin covering the face has a singularly loose and shrivelled appearance and, on a very close inspection, it appears to have been artificially joined to the skin of the head across the eyes and the upper part of the nose. The projections in lieu of ears are composed of folds in the same piece of skin of which the face is formed. The eyes appear to have been distended by some means, so as to have kept very nearly the natural form, and there is a faint appearance as though the cornea had been painted to represent the pupil and iris. The object has been so contrived as to leave no appearance to a cursory observer of its having been opened, but simply dried, and there are two small holes on the fore-

head, through which a string has been passed for its suspension while drying.

There was more. Tufts of black hair had been fixed into each nostril. Beneath the breasts, which seemed to have been padded, was a deep fold camouflaging the join-marks of the upper and lower parts of the body. This lower part was composed of an entire fish – "apparently of the salmon genus" – except for its head, the skin of the fish's back being overlaid onto the orangutan's back. And so forth.

The description given by another observer, Edward Donovan (1768–1837), suggests the exhibit looked rather like the frozen moment in time when a large fish realizes the ape it has been trying to swallow whole is too large for it. All were agreed the "mermaid's" facial expression was one of great distress, as if the creature had died in agony.

If we are to believe the *Autobiography* (1854) of P.T. Barnum (1810–1891), it was this very same specimen that came into his hands in 1842, supposedly brought to New York by a UK naturalist called Dr J. Griffin (in fact a Barnum henchman called Levi Lyman), and, as the "Feejee Mermaid", started the turnstiles lucratively ticking. It, or more likely a substitute, is still preserved.

Eades's/Barnum's mermaid was only one among a profusion that appeared – always for display at a price – during the 19th and early 20th centuries. The last to be given any credence in the West appears to have been one reported in the London *Daily Express* in 1921. Most were very similar; few were of the size – approaching human – one might associate with mermaid legends, most being in the range 40–75cm long. A typical description from an 1881 Boston newspaper betrays much of what was probably the truth:

> This wonder of the deep is in a fine state of preservation. The head and body of a woman are very plainly and distinctly marked. The features of the face, eyes, nose, mouth, teeth, arms, breasts and hair are those of a human being. The hair on its head is of a pale silky blond, several inches in length. The arms terminate in claws closely resembling an eagle's talons instead of fingers with nails. From the waist up, the

> resemblance to a woman is perfect, and from the waist down, the body is exactly the same as of the ordinary mullet of our waters, with its scales, fins and tail perfect.

A very popular type of creature fake, even after the fraudulence of such objects had become widely recognized, was called the basilisk or – once the artifice had been acknowledged – jenny haniver. (The origins of the latter name are unknown.) The basilisk of mythology was about 30cm long, breathed fire, and dealt death from its eyes. These bizarre representations, of similar size or smaller, were constructed by leaving flatfishes such as rays or skates to dry until hard and then carving and twisting the carcase into the desired form. The practice seems ancient, and was probably first inspired by the way the underside of a fish like a ray can resemble a demonic perversion of a human face. Some of these concoctions were remarkable works of craftsmanship – hence their enduring popularity. Modern examples, mostly from Mexico, are crude imitations for the tourist trade.

A rather charming fake was the fur-bearing trout, the trout being indeed a trout but the fur being that of a rabbit. This construction, probably produced by a Canadian taxidermist for fun or as a way of attracting business, is in the Royal Scottish Museum, Edinburgh. Wyoming's equivalent was the jackalope, claimedly a rare species of antlered rabbit but in fact a hare to which had been attached the antlers of an infant pronghorn. In England the practice seems to have been common during the 19th century of killing and stuffing young puppies, then displaying them as adult examples of an exceptionally miniature species. Also from 19th-century England came the "pygmy bison" offered for sale by a Mr Murray of Hastings, who specialized in such curios. Only about 18–20cm tall, this was probably made of, over a wooden frame, the skin of a dog, the hair from a bear, and horns and hooves sculpted from buffalo-horn. And the famous vegetable lamb of Tartary – a plant which, when fully grown, produced a lamblike creature as its fruit – was sufficiently fascinating that as late as 1698 Sir Hans Sloane (1660–1753) was able to display before the Royal Society in London an example that had been sent to him; in fact, as

Sloane correctly surmised, it was a large Chinese fern suitably "improved". Such fakes are still produced in Taiwan for tourists.

Of lesser but still venerable vintage was the great monster of Silver Lake in Wyoming County, New York State. This was first spotted in 1855 by four men and two boys who were out fishing. What they initially thought was a large log floating near their boat proved to be merely the monster's head, so one can only gasp at the true size of the creature. As ever, the initial sighting was followed by an influx of sensation-seekers, who were disappointed, although there were still occasional sightings until 1857, when fire destroyed the Walker House Hotel, in nearby Perry. In the wreckage were discovered the remains of the ingeniously built creature, which had been towed around the lake using ropes, and made to surface and sink through the use of a long hose and a compressed-air supply. A.B. Walker, one of the hoaxers, was the owner of the hotel which, as the nearest to the lake, had been doing a good trade catering for monster-hunters.

Fraudsters have on occasion wandered into classical rather than popular mythology, most notably in the case of the hydra. The Hydra of Greek myth was the nine-headed monster Hercules had to slay as one of his labours: each time he chopped off one of its eight mortal heads (the ninth was immortal) there swiftly grew two replacements.* By the 16th century the business of hydra-fakery was well into its swing. The great Swiss naturalist Konrad von Gesner (1516–1565) wrote of one as if it were genuine, and he was by no means alone. The most famous of all these forgeries probably started innocently enough as a religious statuette of the seven-headed dragon (*Revelation xii*) that heralds the Great Beast 666 (*Revelation xiii*), or even of the Beast itself, which likewise has seven heads. This particular hydra first became known in 1648, when it was plundered from its place on the altar of a church in Prague. Some decades later it arrived in the collection of Johann Anderson, Burgomaster of Hanover. That the creature had seven

---

* The legend is possibly based on the octopus, with its eight tentacles and, for the immortal head, the creature's real one.

The four-legged variety of dragon, according to Athanasius Kircher

rather than the Hydra's nine heads seems to have been no deterrent to enthusiasts. Frederick IV of Denmark (1671–1730) offered 30,000 thalers for it. Albert Seba (1665–1736) featured the beast in the first volume of his vast *Locupletissimi Rerum Naturalium Thesauri Accurata Descriptio* (1734–65).

Possibly the first to expose the imposture was the young Karl von Linné (1707–1778), the naturalist later to become regarded under his romanized name, Carolus Linnaeus, as the founder of modern taxonomy; it is primarily because of his involvement that the whole affair has been so well remembered. He immediately recognized the thing as a fake because God never made any creature with more than one brain. (Linnaeus was on dodgy ground with this dictum, as he would have realized had he known about the larger dinosaurs.) Soon he showed the heads and feet belonged to weasels and the skin was merely snakeskin pasted down firmly over a solid form. As the Burgomaster was trying to sell his specimen at the time and was a powerful man locally, Linnaeus left Hamburg swiftly.

This hydra was built on a much smaller scale than its mythological template; as we noted, faked mermaids were likewise usually miniature. Manifestly, the bigger a beast, the greater the effort and money that had to be put into its construction, and the more likely it was to fall apart before some gullible rich enthusiast had time to be parted from his cash. The same went for "preserved" dragons, which appeared from about the 16th century onwards. Jacob Bobart (1599–1680), Professor of Botany at the University of Oxford, made one as a prank out of a dead rat, and was not altogether delighted when the object began to attain international renown as a genuine relic. Other miniatures tend to be earlier; their smallness was ascribed to their being merely babies. In the collection of curios made by the two Johns Tradescant, father (1570–1638) and son (1608–1662), there was listed a dragon no more than 5cm long, and the largest of a set that went on display in Paris in the mid-16th century was described as wren-sized.

In the mid-18th century a 120cm "dragon" was paraded around England complete with an authoritative description of the way in which, in life, it had mangled its captors with its multiple rows of teeth; it even gained a mention in the *Encyclopedia Britannica*. This impressive creature was constructed from the corpse of an angel shark. Despite its true nature's exposure in print, the "dragon" was still earning money as late as 1861.

Some people really believe species of wild men roam the wastes beyond civilization.

The UK naturalist Charles Waterton (1782–1865) spent the years 1804–24 on extended field trips in the Americas, publishing his recollections in a popular book, *Wanderings in South America, the North-West of the United States, and the Antilles, in the Years 1812, 1816, 1820 and 1824* (1825). The book's frontispiece depicts what appears to be a humanoid head, which he described as that of a Nondescript, a beast whose corpse he had encountered in the course of his travels in Guiana. "In my opinion," wrote Waterton with seeming caution,

his thick coat of hair, and great length of tail, put his species

Charles Waterton's illustration of the Nondescript

out of all question; but then his face and head cause the inspector to pause for a moment, before he ventures to pronounce his opinion of the classification. He was a large animal, and as I was pressed for daylight, and moreover, felt no inclination to have the whole weight of his body upon my back, I contented myself with his head and shoulders, which I cut off; and have brought them with me to Europe. . . .

Some gentlemen of great skill and talent, on inspecting his head, were convinced that the whole series of his features has been changed. Others again have hesitated and betrayed doubts, not being able to make up their minds, whether it be possible that the brute features of the monkey can be changed into the noble countenance of Man.

The stuffed head and shoulders were on display for all to see at Waterton's home, Walton Hall, and were certainly a conversation-piece. So too were other, less controversial, examples of Waterton's taxidermic skills. Some naturalists, notably James Stuart Menteith (1792–1870), went so far as to suggest that, since Waterton was both an expert taxidermist and a renowned practical joker, it was just faintly, ever so faintly possible that . . .

Waterton denied any such aspersions: he claimed he was purposely withholding some of the details concerning his discovery of the Nondescript, and of the taxidermic techniques he had used to preserve it, because he wanted to teach a lesson to the faceless, idiot bureaucrats of the British Treasury, especially a certain J.R. Lushington, who had written to Customs insisting Waterton pay duty on specimens imported from the Americas.

The head was made using that of a red howler monkey. Waterton had removed the skull in its entirety – it being impossible for a monkey's skull to be made to look remotely humanoid – and then treated the skin so it could be moulded into the desired configuration: the features, so it has been said, of the detested Lushington himself.

Sydney Smith (1771–1845), reviewing *Wanderings* in the *Edinburgh Review*, joked:

> In this exhibition [i.e., the frontispiece] our author is surely abusing his stuffing talents, and laughing at the public. It is clearly the head of a Master in Chancery . . .

Waterton himself commented in 1837:

> Some people imagine that I have been guilty of a deception in placing the Nondescript as a frontispiece to the book. Let me assure these worthies that they labour under a gross mistake. I never had the slightest intention to act so dishonourable a part. I purposely involved the frontispiece in mystery . . .

One of the great pioneers of cryptozoology, if one might call him that, was the Scottish judge James Burnett, Lord Monboddo (1714–1799), author of two stupendous six-volume works, *Of the Origin and Progress of Language* (1773–92) and the anonymous *Ancient Metaphysics* (1779–99). Monboddo was certainly an eccentric: he seems to have wished he'd been born in ancient Greece, and he tried to adopt a lifestyle more appropriate to that age than his own. He also had some offbeat theories. He thought human babies were born with tails, which were deftly nipped off by midwives before anyone else could see, and that the higher apes were actually human beings who had yet to master speech. This latter notion was parodied by Thomas Love Peacock (1785–1866) in *Melincourt* (1817), one of whose characters is the MP Sir Oran Haut-ton, captured in childhood in Angola, brought to Britain as a pet/companion, and later educated to the extent he could take a seat in the House. Despite Peacock's mockery, Monboddo had, decades before Darwin, hit on the realization that the great apes are our close kin.

Well, sort of. It was Monboddo's belief, contrary to accepted ideas of the day, that humankind had not been gifted with speech during the Creation but had developed it at the stage when individuals were coming together to form tribes. This happened independently countless times, which is why there are so many different languages. Further, since in some parts of the world the living's so easy people don't need to band together for hunting and defence, there's no reason they should *even yet* have developed speech. And, sure enough, in exactly such areas you find orangutans.

Monboddo had a greatly inflated idea of what orangutans were capable of, and this was because of another theory of his that relates more directly to cryptozoology: that any type of creature which one might imagine *could* exist either does so or has done so. He therefore collected travellers' tales of headless men, mermaids, cyclopes and the like and accepted them as the literal truth. He was, if you like, perpetuating the zoological fantasies of the medieval bestiaries some considerable while after they should have been relegated to the status of curio.

Some of the apparent humans he learned about – mermaids, for example – diverged far further in form from the standard model than did orangutans, so in that context his ideas about the apes seem reasonable. Less so is his notion that – because of their social structures – beavers, too, were latent human beings.

He also believed that the world was nearing its end, this because humans were becoming smaller than in ancient times. (Monboddo himself was only about 1.5m tall, reinforcing his view.) Of course, there was reference in the Bible to days when giants walked the earth, and it seemed to Monboddo that the builders of Stonehenge, and of course his favourite ancient Greeks, must have been several metres tall. There were still, Monboddo learned from travellers and mariners, holdouts of giant people in remoter lands.

Plenty of others aside from Monboddo have been fascinated by the possibility of there being other forms of *Homo* still unknown or unrecognized in the world. Indeed, it seems the urge to discover "lost" forms of humankind is stronger than that for any other type of animal except, in a very different context, the extraterrestrial creature. Maybe in both these instances we're looking for something that'll mirror and thereby cast light on ourselves – the ultimate extension of our tendency to anthropomorphize all the creatures and even machines and inanimate objects around us. Or perhaps this is the same urge as has driven, among others, generations of Leakeys to swelter in the hot African sun seeking our most distant ancestors: manlike cryptozoological creatures like the sasquatch are, in their affect, merely the mobile equivalents of the hard stone fossils the anthropologists unearth. And the fossils have, of course, the advantage that, even if there can be squabbles over matters of interpretation and significance, they're unchallengably *there*.

Fossils, not just of hominids, have attracted their fair share of bogus scientists over the years. The University of Frankfurt's Reiner Protsch von Zieten (b1939) made sensational discoveries in the field of human palaeontology before being exposed in 2004. The Japanese archaeologist Shinichi Fujimura (bc1950), over a career that lasted nearly three decades, unearthed artefacts and other remains that required a rewriting of Japan's prehistory in a manner that appealed to ideas of Japanese nationalism; in 2000 he was exposed as a fraud who'd been planting items for himself later to dig up. Little more than a decade earlier, in 1989, there was the exposure as fraudulent of the Punjabi researcher Viswat Jit Gupta, whose "discoveries" over the preceding quarter-century had required a sensational rewriting of the palaeontological and stratigraphic history of the Himalayas.

The classic example of such shenanigans is Piltdown

19th-century reconstruction of a Neanderthaler, based on fossil evidence

A skull fragment unearthed in 1906 in Nebraska; misinterpretation of this and other remains led to the Nebraska Man fiasco

Man. In 1912 in a gravel near Piltdown Common in Sussex, UK, were discovered a skull and jaw that appeared to be those of a Neanderthal or even pre-Neanderthal hominid. Researches over succeeding years revealed related artefacts;

even when one of these appeared to be a prehistoric cricket bat, the UK anthropological community remained convinced the country had its very own early fossil hominid. The psychic Edgar Cayce was not so easily fooled by these nationalist claims: he was able to determine that *Eanthropus dawsonii*, as Piltdown Man had been named, was actually an Atlantean immigrant, one of many refugees from the foundering continent. Not until 1953 was Piltdown Man revealed as a hoax, the famous fossil head being shown to be part of the skull of a relatively recent human plus the jaw of an orangutan. Debate has raged ever since over the identity of the hoaxer, although the finger is most usually pointed at the man who first discovered the Piltdown assemblage, Charles Dawson (1864–1916), an amateur fossil-hunter and local historian of occasionally question-able ethics.

The UK was not alone in its fossil hominid embarrass-ment: the US had a similar experience with Nebraska Man, and for similar reasons. The important but over-eager palaeontologist at the heart of matters was Henry Fairfield Osborn (1857–1935), Professor of Zoology at Columbia University and Curator of Vertebrate Paleontology at the American Museum of Natural History in New York.

Some unusual bones, including part of a skull, were discovered by journalist Richard Gilder in 1906 near a Native American burial mound outside Omaha, Nebraska. He took them to Erwin Hinkley Barbour (1856–1947), the Nebraska State Geologist, who confirmed their curiousness. Word reached Osborn, who rushed to the area to examine the bones for himself; his examination convinced him these were the earliest known human remains discovered in the Americas, seemingly confirming his own view that human-ity had migrated to the Americas across the Bering land-bridge earlier than generally accepted. A few weeks later he published an article about Nebraska Man in *Century* maga-zine. It was at that point that the Smithsonian sent Ares Hrdlicka (1869–1943) to look the remains over. Hrdlicka correctly inferred they were from a relatively recent Native American burial.

That should have been the end of the story, but in 1922 there was a second hominid flap involving Osborn –

although this time the fault wasn't entirely his. He was sent an apparently anthropoid tooth from Nebraska by an amateur fossil hunter called Harold J. Cook (1887–1962). Osborn was uncertain as to exactly what the tooth was, but thought it might be of a creature related to both primates and humans. He dubbed the hypothetical species *Hesperopithecus*, and talked about it with the anatomist Grafton Elliot Smith (1871–1937), who had been peripherally involved in the Piltdown Man fiasco. Somehow from Smith the tale got to the *Illustrated London News*, which ran a major article heralding *Hesperopithecus* as a great new discovery of prehistoric humans. Osborn wrote a strong letter to Smith, but of course it was too late: it was Nebraska Man all over again. An expedition Osborn sent to Omaha revealed the tooth belonged not to a primate at all but to an extinct peccary. Of course, this announcement wasn't nearly as exciting as the original, and so the legend of Nebraska Man lived on for a while.

Another false start for those seeking to prove people had come to the Americas in very ancient times was Trenton Man. Charles Conrad Abbott (1843–1919) excavated a site near Trenton, New Jersey, in the 1870s, believing there was evidence of prehistoric human activities there, a belief shared by George Frederick Wright (1838–1921), who worked the site from 1880 and wrote two books on the subject, *The Ice Age in North America* (1889) and *Man and the Glacial Period* (1892). In *Glacial Man in Ohio* (1888) Wright understated matters a little:

> In my original "Report upon the Glacial Boundary of Ohio, Indiana and Kentucky," I remarked that since man was in New Jersey before the close of the glacial period, it is also probable that he was on the banks of the Ohio at the same early period; and I asked that the extensive gravel terraces in the southern part of the State be carefully scanned by archeologists, adding that when observers became familiar with the forms of these rude implements they would doubtless find them in abundance. As to the abundance, this prophecy has not been altogether fulfilled.

In 1899 a human femur turned up; unfortunately, it proved

to be fairly recent, and Trenton Man shuffled off a stage he had never really shuffled onto in the first place.

And then there were all the US hoaxes, the most famous being the Cardiff Giant. The hoaxer was a Binghamton, New York State, cigar manufacturer and atheist named George Hull (d1902), who had the idea after arguing with or hearing a tedious sermon by (accounts vary) a Creationist preacher. Hull claimed excavations in Cardiff, New York State, in October 1869 had unearthed the complete fossil of a man – and what a man! Cardiff Man was about 3m tall, and proportioned to match. P.T. Barnum commissioned a duplicate and put it on display to a credulous public only too eager to accept this as a genuine fossil: not only did the US have its own example of a prehistoric hominid, that hominid was *bigger* than anyone else's. By

A contemporary magazine illustration of the Cardiff Giant being "hoisted from its burial place on Newell's Farm"

contrast with the public, palaeontologists recognized from the outset the "fossil" was a fake. In his *Autobiography* (1905) Andrew D. White (1832–1918), co-founder of Cornell University, observed:

> There was ample evidence, to one who had seen much sculpture, that it was carved, and that the man who carved it, though by no means possessed of genius or talent, had seen casts, engravings, or photographs of noted sculptures. The figure, in size, in massiveness, in the drawing up of the limbs and in its roughened surface, vaguely reminded one of Michelangelo's Night and Morning. Of course, the difference between this crude figure and those great Medicean statues was infinite; and yet it seemed to me that the man who had carved this figure must have received a hint from them.

Elsewhere White remarked that there "was very evidently a 'joy in believing' in the marvel" – in other words, that people of all walks *wanted* to believe the Giant really was a fossil hominid.

The object's origins were very soon traced. Hull had commissioned an artist in Fort Dodge, Iowa, to hew a humanoid figure from a block of gypsum he'd purchased; the figure was damaged in transit, and repaired by a Chicago stonecutter. Hull had tried to give the figure an aura of antiquity by treating the surface with sulphuric acid and scuffing it with tools, then taken it to Cardiff for its "discovery". After the exposure, interest in the Giant ebbed fast.

Chagrined to have lost out on the continuing profits from his replica, Barnum commissioned the construction and "discovery" of *another* fake fossil hominid, Colorado Man. This fake was debunked even more swiftly than its predecessor, and Barnum abandoned "fossil" hominids for good.

But others didn't. Further fake human fossils turned up with some regularity in the US through the late 19th and early 20th centuries, including the Taughannock Giant, "discovered" near Ithaca, New York, in 1879. Two metres tall and weighing 450kg, it was made out of iron filings, ox blood, eggs, sand, sugar, salt, sulphur and phosphorus. Its creator, one Ira Dean, having baked his block of this conglomerate, sculpted it with a chisel. Unfortunately, he wasn't able to complete the fake: by the time he got to the statue's feet the mixture had hardened so much that his chisel wouldn't cut into it. Even so, the "fossil" enjoyed a brief vogue.

John G. Eigenmann, a building contractor of Evansville, Indiana, was a colourful character who'd served with distinction in the Civil War. In 1902 he was 65, and his days in the building trade were drawing to a close: a nest-egg for his retirement would be very welcome. Workmen of his, while dredging the River Ohio, found a human body that had apparently "undergone a complete process of petrification". The stone figure was about 1.7m tall, and had been "preserved" with such fidelity that its finger wrinkles and even its ring could be quite clearly seen. Local scholars probed the figure and eventually declared it genuine: apparently the clincher was when they drilled into its head and discovered it was hollow: "The physicians say that if the specimen had been artificially prepared it would have been impossible to make the cranium hollow," reported the *Evansville Courier* for October 4.

Over the next few days, rumours abounded around Evansville as to the possible identity of the petrified man. The idea that he was a murder victim rapidly took hold, and several people claimed him as a lost husband or relative. Meanwhile, a few doubts were surfacing, not least about the fact that the "man" had one leg longer than the other, and similarly one arm more massive than the other. And the local authorities were sceptical too: the Evansville coroner, John P. Walker, declined to perform an inquest – an unusual decision bearing in mind this was a supposed murder victim – while the Evansville Police Chief told reporters that, as far as he was concerned, the figure was a fake; the *Courier*, which leaned Democrat, satirically suggested the petrified man might be a local Republican office-holder, discovered still on the payroll.

Eigenmann put the figure on public display, charging 25 cents a view, and apparently started considering offers to buy it sent by various museums. But then the unexpected dashed his hopes of a retirement fund. One of the men who'd actually pulled the figure from the river bed, Louis Lamb, filed suit, claiming the figure was worth some $7000 and rightfully belonged to him. Before the trial could commence, the stone man mysteriously disappeared. Lamb lost his case, and immediately thereafter Eigenmann too disappeared – next to be heard of a couple of weeks later, now in Rockport, Indiana, and once more exhibiting the "petrified man" for a fee; asked how it could have got to Rockport, he replied, according to the *Evansville Courier* for November 3, that it "had come to life and walked the distance". Seven years later Eigenmann died and the figure was lost.

Fossils, fake or otherwise, are all very well, but what of those reclusive humanoids who're still alive, who leave tracks and scat and are occasionally seen fleetingly on film, but who never seem to oblige us by turning up in person, dead or alive?* In North America they're generally referred to as bigfoot or sasquatch (which are also both often, as here, used as generic terms), in the Himalayan region as the yeti or Abominable Snowman, in Russia, Mongolia and China as the alma; there are countless other local names. A glance at the geographical distribution of bigfoot reports suggests the plausible hypothesis that the creatures might have originated in Asia and many thousands of years ago migrated over the Bering landbridge into Canada and what are now the states of Washington, Oregon and California, and thence into northern South America. This is more or less exactly how *Homo sapiens* spread from the Old World into the Americas.

Assuming these creatures exist, we have no real idea of their zoological standing. Are they all of the same species or closely related species? Are they very large apes or are they more closely related to humankind than to the apes? Are

---

*Or perhaps they do turn up dead. Some enthusiasts suggest bigfoot could be a survival of *Gigantopithecus* (see page 274) or of Neanderthal stock. If so, we'd have to accept there are bigfoot fossils!

Left: From an 18th-century Mongolian manuscript, a drawing of a Gin-Sung – a giant wild man from the Szechuan region

Below: A 1993 bigfoot photo by Ray Wallace (see page 281)

Bottom: The US pioneer hero Andrew Poe (b1742) fights a bigfoot, here identified as a Native American

they walking fossils, showing ourselves as we were millions of years ago? Or do they represent an evolutionary dead end (or ends), an unsuccessful offshoot of the human evolutionary tree that as yet hasn't quite died out? Until we find a corpse or capture a living specimen, all such speculations are fairly redundant.

In 1935 the German palaeontologist Gustav Heinrich Ralph von Koenigswald (1902–1982) discovered in a Hong Kong drugstore a primate molar tooth larger than any hitherto known. (Fossils have a role in traditional Chinese medicine.) He named it *Gigantopithecus blackii* in memory of his friend the anthropologist Davidson Black (1884–1934). Further *Gigantopithecus* fossils have since been found in various parts of Asia, although not as yet in the Americas. It's not inconceivable, although improbable, that the descendants of *Gigantopithecus* could still survive, scattered and few, in the hinterlands of human geographical influence.

It seems to be the idea that bigfoot might be some form of *Homo* rather than a great ape that's the primary reason why so many zoologists are so vociferously antithetical to bigfoot research. If bigfoot were described as merely a large primate, perhaps the zoologists' resistance might be less fierce. It's not as if hitherto undiscovered large primates don't turn up from time to time, so there's certainly precedent for another to be identified. To cite Jeff Meldrum's *Sasquatch: Legend Meets Science* (2006):

> During the past century, over two hundred additional species of primate have been discovered. In the Neotropics alone, twenty-four new species have been described since 1990, and at least ten more await formal description. Most recently, the prospect of a new ape, perhaps something intermediate to a chimp and a gorilla, has sent primatologists converging on the Congo in search of the so-called Bili (or Bondo) ape – with little more evidence to go on than some oversized footprints, nests, a few strands of hair, and persistent native accounts of a large ape, which they call the "lion killer" due to its enormous size.

There seems no perceptible difference between the situation as regards the Bili ape and that as regards the sasquatch, except that the sasquatch is supposedly found in

North America rather than halfway around the globe and the "persistent native accounts" are from Americans (formerly from Native Americans, for centuries before the Europeans arrived) rather than from distant Africans. Oh, and that since Meldrum wrote the above the Bili ape has been found: it's a hitherto unknown giant chimp now allocated the subspecies *Pan troglodytes schweinfurthii*. It seems almost unfair to observe that it has very large feet, of lengths up into the lower range claimed for bigfoot.*

Meldrum, an associate professor of anatomy and anthropology at Idaho State University, is today probably the foremost carrier of the scientific torch for bigfoot research, and a tireless campaigner for the cause that zoology should take bigfoot reports seriously. Aside from bigfoot, his speciality is in bipedal locomotion, and this is particularly valuable when analysing bigfoot tracks. His work has, however, brought him into confrontation with the zoological establishment. Some claim it's not so much Meldrum's bigfoot interest that causes his ostracism, more his aggressive insistence that other zoologists and anthropologists should be taking the subject seriously. Whatever, a comment by New York City anthropologist Todd Disotell resounds: "I think what is happening to him is a shame."†

The 13th-century English scholar Roger Bacon (*c*1214–1294) described how the locals would capture the wild men who inhabited the high mountains of the Far East by setting out dishes of fermented liquor; once the unfortunate yeti had drunk itself into a stupor the hunters would move in. Whether they wanted the captured creature for its flesh or to talk to is unknown to us; Bacon didn't clarify.

---

* Depressingly, the Bili ape may be extinguished before it has properly been found, so to speak. The Bili Forest region is caught up in the midst of the Democratic Republic of Congo's interminable and murderous civil war, and its apes seem to be being used as bargaining chips by the insurgents.

† *Scientific American*, November 19 2007.

The earliest direct encounter between a Westerner and yeti evidence appears to have been in 1889 when Major Laurence Waddell (1854–1938), according to his book *Among the Himalayas* (1899), was travelling in Sikkim when he came across a line of large footprints in the snow. He reported that the Sherpas with him claimed these had been left by giant hairy wild men. By 1921, when a UK expedition led by Colonel Howard Bury (1883–1963) was attempting Everest's north face, interest in the possible existence of a Himalayan wild man was intense, and during the ascent the climbers saw, far ahead on the ice, a number of moving dots. When they reached the place, which was at some 7000m, they found huge footprints. The Brits assumed what they had seen must have been mountain wolves, and that the prints had enlarged through melting in the usual way, but the Sherpas insisted the marks had been left by the *metoh kanqmi* – a term that translates as "abominable snowman".

The expedition's account stimulated other travellers to go public with their own experiences. In an interview with the *Times* for November 2 1921, William Knight (1858–1943) recalled a meeting in the region:

> I stopped to breathe my horse in an open clearing . . . I heard a slight sound, and looking round, I saw some 15 or 20 paces away, a figure which I now suppose must have been one of the hairy men that the Everest Expedition talk about . . . Speaking to the best of my recollection, he was a little under six feet high, almost stark naked in that bitter cold – it was the month of November. He was a kind of pale yellow all over . . . a shock of matted hair on his head, little hair on his face, highly splayed feet, and large, formidable hands. His muscular development in the arms, thighs, legs and chest was terrific. He had in his hand what seemed to be some form of primitive bow.

This seems not at all like the usual "giant hairy apelike creature", and it seems possible Knight's encounter was with one of the religious ascetics who dwell high on the slopes and are said on occasion to walk barefoot, even though the snow and ice must be cruel. The Greek photographer and Fellow of the Royal Geographic Society N.A. Tombazi had a somewhat similar encounter. He concluded in his book *Account of a Photographic Expedition to the Southern Glaciers of Kanqchenjunga in the Sikkim Himalaya* (1925):

> I conjecture then that this "wild man" may be either a solitary or else a member of an isolated community of pious Buddhist ascetics, who have renounced the world and sought their God in the utter desolation of some high place, as yet undesecrated by the world.[*]

Much more recently, Lawrence Swan (d1994), an authority on high-altitude ecology, vehemently disagreed with such explanations:

> The interpretation that tracks in the snow ascribed to the Yeti may be made by man is valid in some instances, but it is clear that footprints cannot logically be attributed to even the most solitary hermit when they are made in remote glaciated terrain at altitudes where local inhabitants simply would not travel.[†]

It is, of course, a long way from saying that tracks could not have been made by humans to insisting they must have been made by the yeti.

The modern Western fascination with the yeti can be said to have got properly under way in 1951 when UK mountaineer Eric Shipton (1907–1977) took photographs in the high Himalayas of mysterious footprints in the snow. In fact, he took two quite distinct sets of photographs of two quite distinct sets of footprints; for a long time the two were considered together, which further served to confuse attempts to determine what creature could have been responsible. In the 1970s it was shown that one track had been left by a quadruped, almost certainly a mountain goat. The other, which shows a five-toed foot, was most definitely not left by a goat and is still puzzling. Because of melting and sublimation, footprints in snow can "grow"; it's not

---

[*] In later life he recanted, saying that with hindsight he thought it must have been a yeti he saw.

[†] Cited in Meldrum's *Sasquatch: Legend Meets Science* (2006).

Supposed yeti footprint photographed by Eric Shipton

impossible the trail was left by one of the religious ascetics mentioned above. We'll likely never know the truth.

In 1970 in Nepal the two UK mountaineers Don Whillans (1933–1985) and Dougal Haston (1940–1977) came across an enigmatic track of footprints at about 4000m. That night, as Whillans looked out of his tent into the moonlight, he saw what seemed to him an ape, moving on all fours. The next day he told the expedition's Sherpas he'd seen a yeti and led them to the footprints he and Haston had discovered. The Sherpas very pointedly ignored the tracks, and Whillans got the impression their attitude was that, if they left the yeti alone, it would leave them alone.

If the relationship between humans and yeti could be characterized as respectful and wary, that between humans and almas, the wild men of northeastern and central Asia, is surprisingly warm: if we are to believe the various accounts, people in those parts treat any alma they come across as you might treat a friendly dog in the street.

An early researcher into the legends concerning Asian wild men was, in the 1880s, Colonel Nikolai Przewalski (1839–1888) of Przewalski's horse fame. In the early 20th century the zoologist V.A. Khaklov collected many accounts; although most of his researches were lost in the chaos of the Revolution, some survived. Reading the descriptions Khaklov recorded in his interviews with Kazakh tribesmen and others, what strikes one immediately is their matter-of-factness. Here's how Khaklov described, based on an interviewee's recollection, a female alma who'd been captured and held for a period of months by local farmers:

> The creature was usually quite silent and bared her teeth on being approached. She had a peculiar way of lying down; she squatted on her knees and elbows, resting her forehead on the ground, and her hands were folded over the back of her head. She would eat only raw meat, some vegetables and gravy, and sometimes insects she caught. When drinking water she would lap, or sometimes dip her arm into the water and then lick her fur.

The other thing that strikes one on reading Khaklov's accounts is that it seems odd – bearing in mind the easy familiarity the Kazakhs he interviewed displayed towards their wild fellows, and the way they seemed to regard meeting almas as a relatively common experience – that Khaklov himself never saw one.

It's a question that could be put to many bigfoot researchers, the world over. One scientist who did report coming face-to-face with an alma was the Leningrad

Drawing by Ivan Sanderson of the female alma described by V.A. Khaklov

University hydrologist Alexander Pronin, who in 1957 was a member of an expedition in the Pamirs. According to a newspaper interview he gave a few months later, he watched this "being of unusual aspect" for several minutes. It was "a manlike creature walking on two feet, slightly stooping, and wearing no clothes. Its thickset body was covered in reddish-grey hair, and it had long arms." His report was not initially given much credit by other Russian scientists, but in due course the Soviet Academy of Sciences set up the Commission for Studying the Question of the Abominable Snowman, with the science historian Boris F. Porchnev (1905–1972) as its head. It was Porchnev's conviction that the yeti and alma alike were survivals of Neanderthal Man, as evidenced by the title of the book he cowrote with the doyen of cryptozoologists Bernard Heuvelmans, *L'Homme de Néanderthal est Toujours Vivant* (1974). Once more, though, the grandiosely named Commission found lots of circum-stantial evidence but no actual yeti/almas, living or dead.

In North America the modern era of sasquatch enthu-siasm began in 1957 when an ex-lumberjack called Albert Ostman claimed that decades earlier, in 1924, he had been held captive for a period of days near Vancouver Island, Canada, by a family of sasquatch – perhaps for breeding purposes, since there was a spinster daughter sasquatch in evidence. Ostman claimed he'd kept silent for over 30 years for fear of ridicule; the obvious retort is that the whole thing was merely a woodsman's tall tale. Nonetheless, in *Sasquatch* Meldrum is much impressed by the fact that Ostman's description of male sasquatch genitalia accords well with the reality of great ape male genitalia, which differ in many respects from their human counterparts; surely, Meldrum seems to be arguing, this is the kind of specialist knowledge you wouldn't expect an ex-lumberjack to have. On the other hand, surely, it's the kind of knowledge an ex-lumberjack could very readily acquire if he wanted to bolster the credi-bility of a tall tale. This is not to say Ostman was lying; simply that one can't use his story, genital expertise notwith-standing, as evidence.

A more significant episode began the following year in Humboldt County, northern California, when the workers on a remote road construction project began complaining that during the nights, when the heavy equipment was left out on its own, sometimes it was being tampered with and frequently, when the men arrived in the mornings, they'd find trails of very large footprints all around. One man, Gerald Crew, made an outline of one of the prints and showed it to local tracker Bob Titmus (d1997). Titmus taught Crew how to cast prints in plaster; it was the first cast Crew made that caught the nation's attention.

The contractor on the project was one Raymond L. Wallace (1918–2002). In later life Ray Wallace became widely known as a bigfoot hoaxer and "entrepreneur", but at the time he seems to have been as baffled as anyone else, and very frustrated by the whole fracas because he missed his deadline for completion of the job and thereby lost money. When Wallace died, nearly half a century later, his surviving family told the media that for years he had been pranking bigfoot trails, and produced the big wooden feet to prove it. Interviewed by the *Seattle Times*, Michael Wallace, Ray's son, went so far as to put it this way: "Ray L. Wallace was Bigfoot. The reality is, Bigfoot just died." The trouble with this scenario is that trackers and zoologists can tell at a glance if footprints have been made by wooden dummies or real feet;* any tracks Wallace had faked this way had instantly been dismissed as hoaxes.

Wallace had also clouded the situation as regards the famous film shot near the construction site but a decade later, in 1967, by bigfoot hunters Roger Patterson and Robert Gimlin (b1932). Stills from this film have become iconic in popular culture. A hairy hominid, obviously female, seemingly taller and bulkier than a human, canters beside a stream and then, turning back for a last look at the camera, vanishes into the trees. The two men produced casts of footprints to back up their claim, and bigfoot devotees maintain the gait of the creature is definitely not human. Even so, sceptics believe all the film shows is a man

---

* As a single example, a wooden foot landing on a stone will push the stone into the earth, while in the same situation with a real foot the flesh will to some extent accommodate the stone's shape.

in a monkey suit. Wallace was fond of intimating slyly that
he'd set up the whole incident, and knew it to be a hoax;
and certainly Wallace made hoax films of his own featuring
people in gorilla suits. This would seem a conclusive
damnation of the Patterson–Gimlin movie were it not that
Wallace's hoax films are transparent frauds whereas the
Patterson–Gimlin movie, if indeed a fake, is a good one.

In 1997 movie director John Landis (b1950) said
Hollywood special-effects expert John Chambers
(1923–2001), who'd been working on *Planet of the Apes*
(1968) at the time, had admitted involvement, but
Chambers himself went to his death denying knowledge of
the incident. In 2004 the writer Greg Long, in his book *The
Making of Bigfoot: The Inside Story*, claimed he'd interviewed
the wearer of the monkey suit, Bob Heironimus, and
learned from him that the suit had been made by one Philip
Morris. It's clear there's really no good explanation of the
Patterson–Gimlin film, although the primatologist John
Napier, among the foremost orthodox scientists seriously to
research bigfoot, was dismissive of it on the grounds that
stride-length and foot-size seemed not to match, that the
creature's buttocks are pronounced like those of a human
rather than flat like those of a great ape, and that the gait
more resembles that of a human taking artificially long
strides than that of a biped taking strides of the length to
which it is accustomed. By contrast Jeff Meldrum, whose
speciality is as we've noted primate locomotion, reckons
there's no problem with the observed gait. Clearly it's a
matter of choosing between specialists.

Patterson himself is manifestly convinced that what he
filmed was what he saw, and various scientific researchers
who have interviewed him are unanimous as to his integrity.
There's a strong school of opinion that Patterson was a
victim of a hoax, and some have pointed the finger at
Gimlin, suggesting he might have been "somewhat of a
'third man' character in this affair", to quote Napier – that
he could have set up the encounter, initially with no grander
an idea than to play a trick on his friend. However, there's
not a shred of evidence, and many independent witnesses
vouch for Gimlin's honesty.

Another episode from the 1960s had the distinction of

involving two highly distinguished cryptozoologists, Bernard Heuvelmans and Ivan T. Sanderson (1911–1973). For some years an individual called Frank K. Hansen had been showing around Midwest carnivals a block of ice containing what was claimed to be a bigfoot-style creature; an alternative explanation was that it could be either a survival or a well preserved fossil of one of our hominid ancestors. It was Hansen's claim (or one of them) that the hairy body had been discovered in a vast block of ice in the Bering Strait.

In late 1968 news of the exhibit reached Sanderson, who tracked it down to a small farm near Winona, Mississippi, where Hansen stored it during the off-season. Heuvelmans, a Belgian, was in the USA at the time, and so Sanderson asked him if he, too, would like to come along and help in the examination of the enigma. By this time it had been realized the body could not be a fossil: fossil humans do not normally have, as cause of death, a bullet-hole in the head. It seemed the relic must be – if anything – a variety of bigfoot.

Heuvelmans and Sanderson spent two days examining the Minnesota Iceman (or "Bozo", as it was also known) and taking photographs. The circumstances were not ideal. Hansen, quite reasonably, would not permit the creature to be thawed out. The body was kept in a plate-glass-topped coffin in a badly lit trailer, and in order to sketch the creature Sanderson had to crawl around on top of the glass. Nonetheless, he and Heuvelmans were sufficiently impressed that they brought out separate scientific papers claiming the creature as a genuine hominid. John Napier, then at the Smithsonian Institution, was on the verge of having a more rigorous examination performed when Hansen's story began to change.

Indeed, it was to change quite frequently hereafter. First it was announced that Hansen was not in fact the Iceman's owner, merely its manager; the real owner was a Hollywood millionaire who wished to remain anonymous. This mystery man had picked up the corpse in Hong Kong, whence it had travelled by a circuitous route to reach the USA. The millionaire was not interested in his "fossil" being subjected to scientific examination: instead, he was determined it be on show so the "common man" could make up his own mind. Hansen, his obedient servant, returned the Iceman to the carnival circuit.

But the Iceman now on display was quite clearly not the one Sanderson and Heuvelmans had examined: for a start, whereas before it had had but a single tooth, now it was the proud owner of four. Hansen admitted only that he had substituted a latex model for the original, presumably to protect the latter from wear and tear.

Then a woman called Helen Westring hit the tabloid headlines with her claim that, a few years earlier, she had been alone in Minnesota woodlands when a great pink-eyed monster with huge hairy hands had raped her; afterwards she had been able to grab her rifle and put a bullet through the rapist's eye. Here was yet another explanation of the beast's origin. Hansen seemed briefly to accept the claim – now depicting himself as the lucky discoverer of the corpse that Westring had left behind.

But then he decided he preferred the limelight exclusively for himself. The "true" story was that, seven or eight years earlier while in the USAF, he had been hunting with companions when he'd startled and shot the wildman. He had left the corpse in the snow for a couple of months before returning with a deep-freeze to reclaim it. Much later he had moved the body to his Winona farm, preparatory to taking it out on tour – which is where it, or its latex replica, still was as late as the 1980s.

The plethora of diverse stories caused Sanderson and Heuvelmans to give up the whole affair as a fraud. Much has since been made of the fact that a number of Hollywood model-makers claimed to have constructed the creature – notably Howard Ball and his son Kenneth – but, since Hansen had stated that it was in Hollywood that he had the *replica* made, these claims don't tell us much about the original.

Anthropological fakery is by no means confined to fossils or bigfoot. The tiny, primitive Tasaday tribe of Mindanao Island in the Philippines first came to world attention in

1971, when an official called Manuel Elizalde (d1997) claimed he'd seen them in the distance years ago while out hunting and had now made proper contact with them. The tribe's cultural level was Palaeolithic: they had not discovered agriculture, had only stone tools, and the few garments they wore were pieced together from leaves. They lived in caves in the middle of the rain forest. Charmingly, they seemed completely without aggression: they had no words for "war" or "weapon", and traded peacefully with the people of other, more advanced cultures around them.

This was a spectacular discovery. The Tasaday were the most primitive known culture in the world. Philippine dictator Ferdinand Marcos promptly declared the region surrounding the Tasaday's caves a no-go area for the public; only anthropologists were allowed in, and then only with the approval and supervision of Elizalde, whom the Tasaday called "*Mono dakel de weta Tasaday*" – "Great man, god of the Tasaday". Various papers on the Tasaday appeared in the anthropological journals, then both a *National Geographic* feature and, shortly afterwards, a CBS television documentary. Elizalde set up a US charity to protect these wonderful people, and drew in large contributions thanks to high-profile sponsors like Charles Lindbergh (1902–1974).

In 1986 Marcos was overthrown, and the democratic regime of Corazon Aquino replaced his dictatorship. One consequence was that access to the Tasaday became freer. The Swiss journalist Oswald Iten (b1950) visited the tribe unannounced and found the caves empty; searching around, he discovered the Tasaday living in modest houses nearby, their garments of leaves replaced by Western attire like jeans. They admitted to him, as he reported in *Neue Zurcher Zeitung*, that the whole affair had been a hoax: they had been paid by Elizalde to act the part of primitives. Whether the scheme was Elizalde's alone is uncertain: he'd certainly made millions out of the US charity he'd founded, and escaped with those millions to Costa Rica after Marcos's downfall, but it's possible also that it was a governmental scam, mounted for political reasons, with Elizalde as just intermediary (and beneficiary). What is even more astonishing than the audacity of the hoax is the gullibility of the anthropologists who, with only a few dissenting sceptics, studied the Tasaday and didn't notice any of the many obvious discrepancies in the whole set-up. As just one example, the bamboo of which their bows and other implements were made cannot grow in the rain forest, but must be specially cultivated elsewhere. As for their being able to survive through food-gathering alone, no anthropologist ever observed them actually gathering any food.

Here's a very partial listing of bigfoot movies in the English language:* *The Geek* (1938), *The Snow Creature* (1954), *Man Beast* (1956), *Abominable Snowman* (1957), *Eegah!* (1962), *Legend of the Blood Mountain* (1965), *Bigfoot* (1967), *Bigfoot* (1970), *Bigfoot: Man or Beast?* (1972), *The Legend of Boggy Creek* (1972),† *Beauties and the Beast* (1974), *Shriek of the Mutilated* (1974), *The Creature from Black Lake* (1976), *Curse of Bigfoot* (1976), *Sasquatch, Legend of Bigfoot* (1976), *Snowbeast* (1977), *Capture of Bigfoot* (1979), *Night of the Demon* (1980), *The Barbaric Beast of Boggy Creek Part II* (1985),‡ *Bigfoot* (1987), *Harry and the Hendersons* (1987), *Demonwarp* (1988), *The Hunt for Bigfoot* (1991), *Bigfoot: Unforgettable Encounter* (1994), *To Catch a Yeti* (1995), *Drawing Flies* (1996), *Little Bigfoot* (1997),§ *Big and Hairy* (1998), *Fear Runs Silent* (1999), *Bigfootville* (2002), *The Untold* (2002), *They Call Him Sasquatch* (2003), *Among Us* (2004), *Suburban Sasquatch* (2004), *Clawed: The Legend of Sasquatch* (2005), *Sasquatch Hunters* (2005), *Abominable* (2006), *The Legend of Sasquatch* (2006), *The Sasquatch Dumpling Gang* (2006), *Sasquatch Mountain* (2006), *Scream of the Sasquatch* (2006), *Yeti: A Love Story* (2006), *Primal* (2007) and *Sludge: The Terror of Bear*

---

* A notable Japanese addition to the canon is *Jû jin yuki otoko* (1955). A crassly edited US version, *Half Human: The Story of the Abominable Snowman*, was released in 1958.

† Plus *Return to Boggy Creek* (1977) and *Boggy Creek 2 – The Legend Continues* (1985).

‡ There appears to have been no *Barbaric Beast of Boggy Creek Part I*.

§ Plus *Little Bigfoot 2 – The Journey Home* (1997).

Claw Mountain (2007). Doubtless you know of others.

It's obvious the high-points of the craze for bigfoot in the movies have been (a) the later 1970s and (b) now. Whether these peaks represent correspon-ding crests in public credulity about bigfoot reports and cryptozoology in general, in the same way that movies about flying saucers flourished during the 1950s and 1960s, is a moot point. However, a couple of recent examples suggest it might be so.

Scientific American, not

The world of bigfoot research was set alight in summer 2008 when former prison officer Rick Dyer and policeman Matt "Gary" Whitton – the latter on sick leave from the Oakland County, Georgia, force after having been shot in the wrist – claimed to have discov-ered a tribe of the creatures living in the mountains of northern Georgia; to prove it they displayed on their recently created website (www.bigfoottracker.com) photos of a large hairy corpse in a freezer. Oh, and on the same website enthusiastic cryptozoologists could sign up for bigfoot expeditions for a mere $499.

The two men also blitzed YouTube with pertinent videos, including an interview with scientist Paul Van Buren, who claimed to have investigated the body and found it genuine. When "Van Buren" was exposed as Whitton's brother, the story shifted: the deception had been perpe-trated in order to throw cryptozoological stalkers off the track. Similarly the story of how the two men acquired the corpse altered: at first they claimed they'd shot the creature but later they amended this, saying instead they'd found it already dead.

By now most of the bigfoot-seeking fraternity had

concluded the men were frauds;* it took the journalistic community a while longer, of course. Since the two men's website advertised that they searched not just for bigfoot and the Loch Ness Monster but also for leprechauns, this is perhaps surprising.

Whitton and Dyer enlisted as their supporter one Tom Biscardi (b1948), a bigfoot-hunter with a reported track record of exaggerated claims; this did nothing to assuage the doubts of the rest of the cryptozoological community but did convince FOX News, who treated him to a long and credulous interview. The photos that had been released on the website a few days earlier, on August 12, had likewise been met with scepticism by researchers, some of whom were quick to point out the surprising resemblance between the corpse in the freezer and the expensive bigfoot costume on sale at www.thehorrordome.com. A news conference on August 15 similarly failed to convince. A couple of days later Steve Kulls of www.squatchdetective.com and Squatch [*sic*] Radio was permitted to examine the supposed corpse as it was thawed out and discovered the exposed head was in places "unusually hollow". The feet proved likewise – indeed, they were made of rubber.

The game up, Whitton and Dyer said it had all been a prank; they had, of course, not had their eyes on any of the various rewards offered by corporate sponsors for conclusive evidence of bigfoot. Whitton was fired from his job as one of Oakland County's finest. Biscardi, despite having claimed earlier to have examined the "corpse" and found it genuine, painted himself as just another victim of the two men's hoax. The latest news is that they're planning to take up hunting for mysterious big cats and dinosaurs instead. Keep an eye open for them buying a Barney costume.

Also in the summer of 2008, hairs supposedly from the Indian variant of the yeti, the mande barung ("forest man"), were tested by scientists first in the UK and then in the US to see what they might be; the hairs had been retrieved by yeti devotee Dipu Marak from a jungle site where a forester had seen a mande barung several times in 2003.

---

* "It just looks like a costume with some fake guts thrown on top for effect," commented Jeff Meldrum.

In the UK in July 2008 a team at Oxford Brookes University led by wildlife biologist and ape conservation expert Ian Redmond and primatologist Anna Nekaris examined the hairs and came to the conclusion they could be from an unknown primate, which might or might not be the fabled yeti. According to Redmond, the hairs were remarkably like some mystery hairs which had been collected by Sir Edmund Hillary during his Himalayan excursions. Unfortunately, when the hairs were sent to the US for DNA testing, they proved to have belonged to a Himalayan goral, a species of goat. Redmond hastily back-tracked: "We always knew that the link between the sightings of the Indian yeti and the finding of the hairs was purely circumstantial." He consoled himself by declaring that at least there seemed to have been a lesser zoological discovery: the geographical range of the Himalayan goral could well be far greater than previously known.

Bigfoot researches can stray out of the cryptozoological into the fanciful. A number of observers have drawn a connection between bigfoot sightings and the appearance of UFOs, as if the creatures might be of extraterrestrial – or even extradimensional – origin. Such an incident is cited in Linda S. Godfrey's curious book *Hunting the American Werewolf* (2006). In 1991 near Marshfield, Wisconsin, one Rita Massman saw a bigfoot near her home on at least four separate occasions and thought this might be linked to the sighting by a local, Robert Kocian, of a UFO while he was out in the early morning cultivating a cornfield.

Godfrey's main concern is the pursuit of a rather different supposed zoological item, the manwolf – a creature, usually moving bipedally, that looks as if it might be a cross between a human or bigfoot and a wolf or dog:

> . . . a creature that stands five to seven feet [1.5–2.1m] tall, covered with dark shaggy fur, that can walk and even run erect yet retains dog-shaped legs and footprints, with a manlike body and a head like a wolf or German shepherd. A long muzzle, fangs, and pointed ears on top of the head complete the picture of one truly unique entity . . .

This is, of course, not the traditional werewolf, the creature that can – or has no choice but to – periodically change its shape between human and wolf form. Even so, in the host of supposed eyewitness accounts that Godfrey breathlessly offers, and most of which yield instantly to interpretation in terms of misperception in conditions of fright and poor lighting, there are several descriptions of frightening beasts morphing between wolf and other shapes, sometimes apelike – hence the potential bigfoot connection – but perhaps more interestingly bearlike. Godfrey links these to a suggestion made in 1974 by Ivan Sanderson that there could still be surviving examples of the Pleistocene wolf *Amphicyon* wandering in the remote wastes of Canada and the northern US. *Amphicyon* had the body of a wolf but was about the size of a grizzly bear and had a much broader face than we associate with a wolf, enhancing the bearlike appearance. It matches the creature called the waheela found in the legends of the Native Americans of northern Michigan. In a similar category is the massive wolf *Canis dirus*, the dire wolf, supposedly extinct but who can say for sure?

In fact, although we can't be certain about *Canis dirus* and *Amphicyon* and for that matter *Gigantopithecus*, we can lay some fairly safe bets. It is not impossible that survivals of these creatures may still roam wilderness regions, just as it's not impossible that plesiosaurs may lurk in the depths of lakes like Loch Ness, but the probabilities are not in favour. It is very nearly beyond the bounds of explicability that, if these creatures still exist, all the years of sometimes obsessive searching for them have yet to turn up a specimen.

On the other hand, the existence of the Komodo dragon – a huge carnivorous lizard, *Varanus komodoensis*, native to four remote Indonesian islands and the nearest thing to a surviving dinosaur you're ever likely to find – was not confirmed until 1910.

Select Bibliography

Babson, Roger W.: *Actions and Reactions: An Autobiography*, Harper & Brothers, New York, 1950 (2nd rev edn of a 1935 book)

Baring-Gould, Sabine: *The Book of Were-wolves, Being an Account of a Terrible Superstition*, Smith, Elder & Co., London, 1865

Bearden, T.E.: *Solutions to Tesla's Secrets and the Soviet Tesla Weapons*, Tesla Book Company, Millbrae CA, 1981

Bell, James Scott: *The Darwin Conspiracy*, Vision House, Gresham OR, 1995

Berlitz, Charles: *Atlantis – The Eighth Continent*, Putnam, New York, 1984

Bouw, Gerardus D.: *A Geocentricity Primer: An Introduction to Biblical Cosmology*, The Biblical Astronomer, Cleveland OH, 2004 (rev edn of 1999 book)

Bouw, Gerardus D.: *With Every Wind of Doctrine: Biblical, Historical, and Scientific Perspectives on Geocentricity*, Tychonian Society, Cleveland OH, 1984

Browne, Sylvia: *Secrets & Mysteries of the World*, Hay House, Carlsbad CA, 2005

Bulgatz, Joseph: *Ponzi Schemes, Invaders from Mars & More Extraordinary Popular Delusions and the Madness of Crowds*, Harmony, New York, 1992

Cayce, Edgar Evans: *Edgar Cayce on Atlantis*, Howard Baker, London, 1969

Dean, Jodi: *Aliens in America: Conspiracy Cultures from Outerspace to Cyberspace*, Cornell University Press, Ithaca NY, 1998

de Camp, L. Sprague: *The Ragged Edge of Science*, Owlswick, Philadelphia, 1980

De Morgan, Augustus: *A Budget of Paradoxes*, ed David Eugene Smith, Open Court, Chicago, 1915 (rev edn of 1872 book)

Dircks, Henry: *Perpetuum Mobile, or A History of the Search for Self-Motive Power, from the 13th to the 19th Century*, Spon, London 1870

Donnelly, Ignatius L.: *Atlantis – The Antediluvian World*, Harper, New York, 1882

Dunne, J.W.: *An Experiment with Time*, Faber, London, 1939 (5th edn)

Dunne, J.W.: *The Serial Universe*, Faber, London, 1934

Evans, Christopher: *Cults of Unreason*, Harrap, London, 1973

Fagan, Garrett G. (ed): *Archaeological Fantasies: How Pseudoarchaeology Misrepresents the Past and Misleads the Public*, Routledge, Abingdon, 2006

Feder, Kenneth L.: *Frauds, Myths, and Mysteries: Science and Pseudoscience in Archaeology*, Mayfield, Mountain View CA, 1999 (3rd edn)

Fort, Charles: *Lo!*, Claude Kendall, New York, 1931

Fort, Charles: *New Lands*, Boni & Liveright, New York, 1923

Fort, Charles: *Wild Talents*, Claude Kendall, New York, 1932

Fort, Charles: *The Book of the Damned*, Liveright, New York, 1919

Franke, Thorwald C.: "King Italos = King Atlas of Atlantis? A Contribution to the Sea Peoples Hypothesis", Proceedings of the 2008 Atlantis Conference, 2008

Friedlander, Michael W.: *At the Fringes of Science*, Westview, Boulder CO, 1995

Gardner, Martin: *The New Age: Notes of a Fringe Watcher*, Prometheus, Buffalo NY, 1988

Garwood, Christine: *Flat Earth: The History of an Infamous Idea*, Thomas Dunne Books, New York, 2008

Godfrey, Linda S.: *Hunting the American Werewolf: Beast Men in Wisconsin and Beyond*, Trails, Madison WI, 2006

Grant, John: *A Directory of Discarded Ideas*, Ashgrove, Sevenoaks, 1981

Grant, John: *Dreamers: A Geography of Dreamland*, Ashgrove, Bath, 1984

Greer, Steven M., and Loder, Theodore C. III: *Disclosure Project Briefing Document*, Disclosure Project, Largo MD, 2001

Hancock, Graham: *Underworld: The Mysterious Origins of Civilization*, Crown, New York, 2002

Hanson, James: *A New Interest in Geocentricity*, Bible-Science Association, Minneapolis, 1979

Henry, William: *One Foot in Atlantis: The Secret Occult History of World War II and its Impact on New Age Politics*, Earthpulse Press, Anchorage, 1998

Hiscox, Gardner M.: *Mechanical Appliances, Mechanical Movements, and Novelties of Construction*, Norman W. Henley, New York, 1904

Huston, Peter: *Scams from the Great Beyond*, Paladin, Boulder CO, 1997

Jastrow, Joseph: *Error and Eccentricity in Human Belief* (retitled reissue of *Wish and Wisdom: Episodes in the Vagaries of Belief*, 1935), Dover, New York, 1962

Jastrow, Joseph: *Fact and Fable in Psychology*, Houghton Mifflin, Boston, 1900

Jordan, David Starr: *The Higher Foolishness, with Hints as to the Care & Culture of Aristocracy, Followed by Brief Sketches on Ecclesiasticism, Science & the Unfathomed Universe*, Bobbs-Merrill, Indianapolis, 1927

Jordan, David Starr: *The Stability of Truth: A Discussion of Reality as Related to Thought and Action*, Holt, New York, 1911

Kenyon, J. Douglas: *Forbidden Science: From Ancient Technologies to Free Energy*, Bear & Co., Rochester VT, 2008

King, David: *Finding Atlantis: A True Story of Genius, Madness, and an Extraordinary Quest for a Lost World*, Harmony, New York, 2005

Knight, Damon: *Charles Fort: Prophet of the Unexplained*, Doubleday, New York, 1970

Kusche, Larry: *The Bermuda Triangle Mystery – Solved*, Prometheus, Buffalo NY, 1986 (rev edn of 1976 book)

Kusche, Larry: "Critical Reading, Careful Writing, and the Bermuda Triangle", *Skeptical Inquirer*, Fall 1977

Leonard, George H.: *Somebody Else is on the Moon*, McKay, New York, 1976

Lépine, François: *Quantum Buddhism: Mahajrya Bodhana Sutra/Teachings on Awakening to the Great Field*, F. Lépine Publishing, St-Raymond, Quebec, 2008

Ley, Willy: *Another Look at Atlantis*, Doubleday, New York, 1969

Livingston, James D.: "Magnetic Therapy: Plausible Attraction?", *Skeptical Inquirer*, July 1998

Lyne, William: *Occult Science Dictatorship: The Official State Science Religion and How to Get Excommunicated – A Book about Alternate Science, Free Energy, UFOs and Government Thought Control*, Creatopia, Lamy NM, 2002

McTaggart, John McTaggart Ellis: "The Unreality of Time", in *Mind: A Quarterly Review of Psychology and Philosophy* #17, 1908

Meldrum, Jeff: *Sasquatch: Legend Meets Science*, Forge, New York, 2006

Michell, John: *Eccentric Lives and Peculiar Notions*, Thames & Hudson, London, 1984

Millard, Joseph: *Edgar Cayce, Mystery Man of Miracles*, Fawcett, Greenwich CN, 1967 (rev edn of 1956 book)

Mooallem, Jon: "A Curious Attraction: On the Quest for Antigravity", *Harper's Magazine*, October 2007

Napier, John: *Bigfoot: The Yeti and Sasquatch in Myth and Reality*, Cape, London, 1972

Nichelson, Oliver: "Nikola Tesla's Later Energy Generation Designs", paper presented to the 26th Intersociety Energy Conversion Engineering Conference, 1991, reprinted in the same author's *Tesla's Fuelless Generator and Wireless Method: Analytical Papers*, 2007

Niemitz, Hans-Ulrich: "Did the Early Middle Ages Really Exist?", apparently online publication only, 1995 (rev 1997, rev 2000)

O'Murchu, Diarmuid: *Quantum Theology: Spiritual Implications of the New Physics*, Crossroad, New York, 1997

Ord-Hume, Arthur W.J.G.: *Perpetual Motion: The History of an Obsession*, St Martin's, New York, 1977

Pickover, Clifford A.: *Strange Brains and Genius: The Secret Lives of Eccentric Scientists and Madmen*, Plenum, New York, 1998

Platt, Charles: "Breaking the Law of Gravity", *Wired* #6.03, 1998

Poythress, Vern Sheridan: "Why Scientists Must Believe in God: Divine Attributes of Scientific Law", *Journal of the Evangelical Theological Society*, March 2003

Priestley, J.B.: *Man and Time*, Aldus, London, 1964

Radner, Daisie, and Radner, Michael: *Science and Unreason*, Wadsworth, Belmont CA, 1982

Randi, James: *Flim-Flam!: The Truth about Unicorns, Parapsychology, and Other Delusions*, Lippincott & Crowell, New York, 1980

Regal, Brian: *Human Evolution: A Guide to the Debates*, Santa Barbara, ABC–CLIO, 2004

Schadewald, Robert J.: "The Perpetual Quest", in *The Fringes of Reason: A Whole Earth Catalog*, ed Ted Schultz, Harmony Books, New York, 1989

Schadewald, Robert J. (ed Lois A. Schadewald): *Worlds of Their Own: A Brief History of Misguided Ideas: Creationism, Flat-Earthism, Energy Scams, and the Velikovsky Affair*, Xlibris, Philadelphia, 2008

Schultz, Ted (ed): *The Fringes of Reason: A Field Guide to New Age Frontiers, Unusual Beliefs and Eccentric Sciences*, Harmony, New York, 1989

Shuker, Karl P.N.: *Mystery Cats of the World*, Hale, London, 1989

Simanek, Donald E.: "Perpetual Futility: A Short History of the Search for Perpetual Motion", apparently online publication only, 2007

Simanek, Donald E.: "Turning the Universe Inside-Out: Ulysses Grant Morrow's Naples Experiment", apparently online publication only, 2003

Stearn, Jess: *Edgar Cayce: The Sleeping Prophet*, Doubleday, Garden City NJ, 1967

Thatcher, Oliver J. (ed): *The Library of Original Sources*, University Research Extension Co., Milwaukee WI, 1901

Tsarion, Michael: *Atlantis, Alien Visitation, and Genetic Manipulation*, Angels at Work Publishing, Santa Clara CA, 2002

van der Kamp, Walter: *De Labore Solis: Airy's Failure Reconsidered*, self-published, Pitt Meadows, British Columbia, 1988

van der Kamp, Walter: *The Whys and Wherefores of Geocentrism*, in *Bulletin of the Tychonian Society* #49–#52, 1988–9

Verance, Percy: *Perpetual Motion, Comprising a History of the Efforts to Attain Self-Motive Mechanism with a Classified, ILLUSTRATED Collection and Explanation of the Devices Whereby It Has Been Sought and Why They Failed*, 20th Century Enlightenment Specialty Co., Chicago, 1916

Wallace, Alfred Russel: *My Life: A Record of Events and Opinions*, Chapman & Hall, London, 1905

Wells, Carol G.: *Right Brain Sex: Using Creative Visualization to Enhance Sexual Pleasure*, Prentice Hall, New York, 1989

Wicker, Christine: *Not in Kansas Anymore: A Curious Tale of how Magic is Transforming America*, HarperSanFrancisco, 2005

Wilson, Colin: *From Atlantis to the Sphinx*, Fromm International, New York, 1996

Wilson, Colin: *Rogue Messiahs: Tales of Self-Proclaimed Saviors*, Hampton Roads, Charlottesville VA, 2000

Wilson, Colin, and Evans, Christopher (eds): *The Book of Great Mysteries*, Dorset, New York, 1990

Wilson, Colin, and Flem-Ath, Rand: *The Atlantis Blueprint: Unlocking the Ancient Mysteries of a Long-Lost Civilization*, Delacorte, New York, 2001

INDEX